T0185918

SpringerBriefs in History of Science and Technology

More information about this series at http://www.springer.com/series/10085

Walter Dittrich

Reassessing Riemann's Paper

On the Number of Primes Less Than a Given Magnitude

Second Edition

 Springer

Walter Dittrich
Institut für Theoretische Physik
Universität Tübingen
Tübingen, Germany

ISSN 2211-4564　　　　　　　　ISSN 2211-4572　(electronic)
SpringerBriefs in History of Science and Technology
ISBN 978-3-030-61048-7　　　　ISBN 978-3-030-61049-4　(eBook)
https://doi.org/10.1007/978-3-030-61049-4

Georg Friedrich Bernhard Riemann (1826 – 1866)

Preface to the Second Edition

In the first edition, I concentrated mainly on the historical development of the Riemann zeta function and its application in mathematics. In this second edition, I have added three new chapters in order to underline the importance of the zeta-function regularization (ZFR) for physics, with the intent to emphasize Riemann's zeta function as a powerful tool to regularize the otherwise infinite quantities that occur in many problems in quantum mechanics and quantum field theory.

The calculations of the partition function of the Bose oscillator in Chap. 9 and of the Fermi oscillator in Chap. 10 are not only challenging but also represent historically the first use of ZFR by Gibbons and Hawking in the seventies.

Shortly thereafter, I proved the usefulness of ZFR in quantum electrodynamics (QED) (Chap. 11). Equation (11.33) represents the result. The path I took is described in great detail, where I show the computation of the one-loop effective action in spinor QED using Riemann's ZFR.

Finally, I thought it would be helpful to the reader to summarize the results of the Euler-Riemann equation and include many findings and graphic representations.

I have also re-written Appendix A.2, improving on the formulation in the first edition.

Tübingen, Germany   Walter Dittrich

Preface to the First Edition

This book is devoted to one of the members of the Göttingen triumvirate, Gauß, Dirichlet, and Riemann. It is the latter to whom I wish to pay tribute, and especially to his world-famous article of 1859, which he presented in person at the Berlin Academy upon his election as a corresponding member. His article entitled, "Über die Anzahl der Primzahlen unter einer gegebenen Größe" ("On the Number of Primes Less Than a Given Magnitude"), revolutionized mathematics worldwide. Included in the present book is a detailed analysis of Riemann's article, including such novel concepts as analytical continuation in the complex plane; the product formula for entire functions; and, last but not least, a detailed study of the zeros of the so-called Riemann zeta function and its close relation to determining the number of primes up to a given magnitude, i.e., an explicit formula for the prime number counting function.

Tübingen, Germany

Walter Dittrich

Contents

Chapter 1
Towards Euler's Product Formula and Riemann's Extension of the Zeta Function

There is a very close connection between the sums of the reciprocals of the integers raised to a variable power that Euler wrote down in 1737, the now-called zeta function,

$$\zeta(s) = \sum_{n=1}^{\infty} \frac{1}{n^s} = 1 + \frac{1}{2^s} + \frac{1}{3^s} + \frac{1}{4^s} + \frac{1}{5^s} + \cdots, \qquad s > 1 \qquad (1.1)$$

and the primes—which, as integers, are the very signature of discontinuity. Euler considered s to be a real integer variable with $s > 1$ to insure convergence of the sum. Multiplying the definition of $\zeta(s)$ by $1/2^s$ we obtain

$$\frac{1}{2^s}\zeta(s) = \sum_{n=1}^{\infty} \frac{1}{(2n)^s} = \frac{1}{2^s} + \frac{1}{4^s} + \frac{1}{6^s} + \frac{1}{8^s} + \cdots \qquad (1.2)$$

and subtracting this from $\zeta(s)$ we get

$$\zeta(s) - \frac{1}{2^s}\zeta(s) = \sum_{n=1}^{\infty} \frac{1}{n^s} - \sum_{n=1}^{\infty} \frac{1}{(2n)^s}$$

or
$$\left(1 - \frac{1}{2^s}\right)\zeta(s) = 1 + \frac{1}{3^s} + \frac{1}{5^s} + \frac{1}{7^s} + \frac{1}{9^s} + \frac{1}{11^s} + \cdots . \qquad (1.3)$$

Hence all the multiples of the prime $n = 2$ disappeared from the original sum of the defined $\zeta(s)$. In short, we found

$$\left(1 - \frac{1}{2^s}\right)\zeta(s) = \sum_{\substack{n=1 \\ n \neq 2k}}^{\infty} \frac{1}{n^s}. \qquad (1.4)$$

© The Author(s), under exclusive license to Springer Nature Switzerland AG 2021
W. Dittrich, *Reassessing Riemann's Paper*,
SpringerBriefs in History of Science and Technology,
https://doi.org/10.1007/978-3-030-61049-4_1

Next, we multiply this last result by $1/3^s$ to obtain

$$\frac{1}{3^s}\left(1-\frac{1}{2^s}\right)\zeta(s) = \sum_{\substack{n=1 \\ n\neq 2k}}^{\infty}\frac{1}{(3n)^s} = 1+\frac{1}{3^s}+\frac{1}{9^s}+\frac{1}{15^s}+\frac{1}{21^s}+\cdots \qquad (1.5)$$

and so, subtracting this from $(1-1/2^s)\zeta(s)$, we have

Leonhard Euler (1707 – 1783);
Drawing by C.F. Gauß

$$\left(1 - \frac{1}{2^s}\right)\left(1 - \frac{1}{3^s}\right)\zeta(s) = 1 + \frac{1}{5^s} + \frac{1}{7^s} + \frac{1}{11^s} + \cdots$$

$$= \sum_{\substack{n=1 \\ n \neq 2k \\ n \neq 3k}}^{\infty} \frac{1}{n^s}. \tag{1.6}$$

Now we multiply this result by $1/5^s$ and so on. As we repeat this process over and over, multiplying through our last result by $1/p^s$, where p denotes successive primes, we subtract out all the multiples of the primes. Hence, since all integers are composed of primes (Euclid's fundamental theorem of the theory of numbers), we removed all numbers of the right-hand side of the defining sum of $\zeta(s)$—except for the number 1. Thus our final result is the product

$$\left\{\Pi_{p\,\text{prime}}\left(1 - p^{-s}\right)\right\}\zeta(s) = 1 \tag{1.7}$$

or

$$\boxed{\zeta(s) = \Pi_{p\,\text{prime}}\frac{1}{1 - p^{-s}} = \sum_{n=1}^{\infty}\frac{1}{n^s}, \quad s > 1}. \tag{1.8}$$

Euler's actual statement reads: "Si ex serie numerorum primorum sequens formetur expressio $\prod_{p}\frac{p^s}{(p^s-1)}$ erit eius valor aequalis summae huius seriei $\sum_{n=1}^{\infty}\frac{1}{n^s}$."

Now we are going to extend Euler's zeta function into the complex plane C, which is a major achievement of Riemann's. Hence from now on, s is complex valued and we write

$$\zeta(s) = \sum_{n=1}^{\infty}\frac{1}{n^s} = \frac{1}{1^s} + \frac{1}{2^s} + \frac{1}{3^s} + \cdots \quad \text{but with} \quad Re(s) > 1. \tag{1.9}$$

This is an absolutely convergent infinite series, which also holds true for the product of all primes in

$$\zeta(s) = \Pi_{p\,\text{prime}}\frac{1}{1 - p^{-s}} = \left(\frac{1}{1 - 2^{-s}}\right)\cdot\left(\frac{1}{1 - 3^{-s}}\right)\cdot\left(\frac{1}{1 - 5^{-s}}\right)\cdots\left(\frac{1}{1 - p^{-s}}\right)\cdots. \tag{1.10}$$

$\zeta(s)$ has no zeros in the region $Re(s) > 1$, as none of these factors have zeros. However, with Riemann's extension of zeta into the entire complex plane, we will be able to locate zeros as well as poles. To show this, we have to analytically continue Euler's original real valued zeta function into the entire complex s plane. A first result in this direction will be achieved with the aid of the so-called Dirichlet series, which turns up when calculating

$$(1 - 2^{1-s})\zeta(s) = \sum_{n=1}^{\infty} n^{-s} - 2^{1-s} \sum_{n=1}^{\infty} n^{-s} = \sum_{n=1}^{\infty} n^{-s} - 2\sum_{n=1}^{\infty} (2n)^{-s}$$

$$= 1 - \frac{2}{2^s} + \frac{1}{2^s} - \frac{2}{4^s} + \cdots = 1 - \frac{1}{2^s} + \frac{1}{3^s} - \frac{1}{4^s} + \frac{1}{5^s} - \frac{1}{6^s} + \cdots$$

$$= \sum_{n=1}^{\infty} \frac{(-1)^{n+1}}{n^s} =: \eta(s), \qquad \text{Dirichlet series}. \tag{1.11}$$

Gustav Lejeune Dirichlet (1805 – 1859)

This series is convergent for all $s \in C$ with $Re(s) > 0$. Hence we can define

$$\zeta(s) = \frac{1}{1 - 2^{1-s}} \sum_{n=1}^{\infty} \frac{(-1)^{n+1}}{n^s} \quad \text{for} \quad Re(s) > 0 \quad \text{and} \quad 1 - 2^{1-s} \neq 0. \tag{1.12}$$

When we write

$$\eta(s) + \frac{2}{2^s}\zeta(s) = \sum_{n=1}^{\infty} \frac{(-1)^{n+1}}{n^s} + \frac{2}{2^s}\sum_{n=1}^{\infty}\frac{1}{n^s}$$

$$= \sum_{n=1}^{\infty}\left(\frac{1}{(2n-1)^s} - \frac{1}{(2n)^s} + \frac{2}{(2n)^s}\right)$$

$$= \sum_{n=1}^{\infty}\frac{1}{n^s} = \zeta(s), \tag{1.13}$$

we can collect our results so far in the string of formulae

$$\zeta(s) = \frac{1}{1-2^{1-s}}\sum_{n=1}^{\infty}\frac{(-1)^{n+1}}{n^s} = \frac{\eta(s)}{1-2^{1-s}} = \frac{1}{s-1}\sum_{n=1}^{\infty}\left(\frac{n}{(n+1)^s} - \frac{n-s}{n^s}\right). \tag{1.14}$$

Most important, we can continue $\zeta(s)$ into the realm of the critical strip $0 < Re(s) < 1$. Of course, the zeros in the denominator in the representation given above have to be excluded, i.e., from

$$1 - 2^{1-s} = 0 \tag{1.15}$$

follows

$$1 = e^{(1-s)\log 2} \tag{1.16}$$

meaning

$$2\pi i n = (1-s)\log 2 \tag{1.17}$$

or

$$s = 1 - \frac{2\pi i n}{\log 2}, \qquad n \in \mathbb{Z}. \tag{1.18}$$

Having shown that the zeta function can be analytically continued into the half plane $\{s \in C | Re(s) > 0, s \neq 1\}$, we still have to prove that $\zeta(s)$ has a pole at $s = 1$:

$$\lim_{s\to 1}(s-1)\zeta(s) = \lim_{s\to 1}\frac{(s-1)}{1-2^{1-s}}\sum_{n=1}^{\infty}(-1)^{n+1}n^{-s} = \lim_{s\to 1}\frac{(s-1)}{1-2^{1-s}}\log 2$$

$$= \lim_{s\to 1}\frac{1}{-\log 2 \cdot 2^{1-s}\cdot(-1)}\log 2 = \lim_{s\to 1}\frac{1}{2^{1-s}} = 1, \tag{1.19}$$

where we used Abel's theorem $\lim_{x\to 1^-}\log(x+1) = \log 2$ and the continuity of $\log(x+1)$. How about arguments for the zeta function equal to or less than zero? Later we will show that the zeta function satisfies the functional equation

$$\zeta(s) = 2^s \pi^{s-1} \sin\left(\frac{\pi s}{2}\right) \Gamma(1-s)\zeta(1-s). \tag{1.20}$$

This defines $\zeta(s)$ in the whole complex s plane. Note that the left-hand side goes over by just changing $s \to 1-s$ into $\zeta(1-s)$, so that we can compute $\zeta(1-s)$, given $\zeta(s)$, e.g.., $\zeta(-15)$ in terms of $\zeta(16)$.

If s is a negative even integer, then $\zeta(s) = 0$ because the factor $\sin(\pi s/2)$ vanishes. These are the trivial zeros of the zeta function. So all non-trivial zeros lie in the critical strip where s has a real part between 0 and 1.

Here is a first curiosity that needs further interpretation. If one substitutes in the functional equation $s = -1$, one obtains

$$\zeta(-1) = 2^{-1}\pi^{-2}(-1)\Gamma(2)\zeta(2) = \frac{1}{2} \cdot \frac{1}{\pi^2}(-1) \cdot 1 \cdot \frac{\pi^2}{6} = -\frac{1}{12}, \tag{1.21}$$

which means that $\zeta(-1) = -1/12$.

This regularized value of $\zeta(-1)$ has absolutely nothing to do with the real-space representation of $\zeta(-1)$ by the divergent series $\zeta(-1) = \sum_{n=1}^{\infty} \frac{1}{n^{-1}} = 1 + 2 + 3 + 4 + \cdots$, which tells us that the same function can have different representations. Some very learned mathematicians entertain the opinion that the zeta-function regularization has swept away the ugly infinities and produced the "golden nugget" of the otherwise nonconvergent series. In quantum field theory one observes the same phenomena, where the zeta-function regularization makes infinities disappear (Casimir effect, quantum electrodynamics, quantum chromodynamics and particle production near black holes). We will come back to this point toward the end of this book.

Chapter 2
Prime Number Counting Function

On the way to showing the significance of the zeta zeros for counting prime numbers up to a given magnitude, Riemann introduces an important weighted prime number function $f(x)$. We will call it $\Pi(x)$ while others use $J(x)$. Since this function is of utmost importance, we will start introducing it by way of examples.

First, the definition of $\Pi(x)$ is given by

$$\Pi(x) = \sum_{\substack{p^n < x \\ p\,\text{prime}}} \frac{1}{n}, \tag{2.1}$$

i.e., for every prime number power p^n which is smaller than x, we sum up its fractions; for example,

$$\Pi(20) = \left(\frac{1}{1} + \frac{1}{2} + \frac{1}{3} + \frac{1}{4}\right) + \left(\frac{1}{1} + \frac{1}{2}\right) + \left(\frac{1}{1}\right) + \left(\frac{1}{1}\right) + \left(\frac{1}{1}\right) + \left(\frac{1}{1}\right)$$

$$\underset{2^1, 2^2, 2^3, 2^4 < 20}{} \quad \underset{3^1, 3^2 < 20}{} \quad \underset{5^1 < 20}{} \quad \underset{7^1 < 20}{} \quad \underset{11^1 < 20}{} \quad \underset{13^1 < 20}{}$$

$$+ \left(\frac{1}{1}\right) + \left(\frac{1}{1}\right)$$

$$\underset{17^1 < 20 \quad 19^1 < 20}{} \tag{2.2}$$

The brackets can also be reorganized like this:

$$\Pi(20) = \left(\frac{1}{1} + \frac{1}{1} + \frac{1}{1} + \frac{1}{1} + \frac{1}{1} + \frac{1}{1} + \frac{1}{1} + \frac{1}{1}\right)$$

$$+ \frac{1}{2}\left(\frac{1}{1} + \frac{1}{1}\right) + \frac{1}{3}\left(\frac{1}{1}\right) + \frac{1}{4}\left(\frac{1}{1}\right). \tag{2.3}$$

© The Author(s), under exclusive license to Springer Nature Switzerland AG 2021
W. Dittrich, *Reassessing Riemann's Paper*,
SpringerBriefs in History of Science and Technology,
https://doi.org/10.1007/978-3-030-61049-4_2

The first pair of brackets counts the number of primes smaller than $x = 20$; the second pair counts the primes that are smaller than the square root of x, etc. Hence, denoting the number of primes up to x by $\Pi(x)$, we get Riemann's formula,

$$\Pi(x) = \sum_{n=1}^{\infty} \frac{1}{n} \pi(x^{1/n}), \tag{2.4}$$

which contains a finite number of terms, which becomes evident by looking at the following example:

$$\Pi(x) = \pi(x) + \frac{1}{2}\pi(\sqrt[2]{x}) + \frac{1}{3}\pi(\sqrt[3]{x}) + \frac{1}{4}\pi(\sqrt[4]{x}) + \cdots$$

$$x = 100:$$

$$\sqrt[2]{x} = 10,\ \sqrt[3]{x} = 4.6415,\ \sqrt[4]{x} = 3.1622,\ \sqrt[5]{x} = 2.51188,$$

$$\sqrt[6]{x} = 2.15\ldots,\ \sqrt[7]{x} = 1.930\ldots < 2. \tag{2.5}$$

If the argument of Π is less than 2, then $\Pi(x) = 0$. So our result for $\Pi(100)$ is given by

$$\Pi(100) = \pi(100) + \frac{1}{2}\pi(10) + \frac{1}{3}\pi(4.6415) + \frac{1}{4}\pi(3.1622)$$

$$+ \frac{1}{5}\pi(2.5118) + \frac{1}{6}\pi(2.15) + 0 + 0 + \cdots. \tag{2.6}$$

Counting the primes, we obtain

$$\Pi(100) = 25 + \frac{1}{2} \cdot 4 + \frac{1}{3} \cdot 2 + \frac{1}{4} \cdot 2 + \frac{1}{5} \cdot 1 + \frac{1}{6} \cdot 1$$

$$= 28\frac{8}{15} = 28.533. \tag{2.7}$$

Hence, for any argument $x > 1$, the value $\Pi(x)$ can be worked out for a finite sum. So far we have learned that $\Pi(x)$ measures primes. Evidently $\Pi(x)$ is a step function which starts at $\Pi(0) = 0$ and jumps at positive integers, i.e., the jump is 1 at primes, 1/2 at squares of primes, and 1/3 at cubes of primes. Hence, our defining equations for $\Pi(x)$ can also be written as

$$\Pi(x) = \sum_{p} \sum_{n=1}^{\infty} \frac{1}{n}\Theta(x - p^n), \tag{2.8}$$

where $\Theta(x)$ is the Heaviside step function given by $\Theta(x) = \begin{cases} 1,\ x > 0 \\ \frac{1}{2},\ x = 0 \\ 0,\ x < 0 \end{cases}$.

There is still another function of the analytical theory of numbers which we need. It is the so-called Möbius function, which defines the inverse of the zeta function:

$$\frac{1}{\zeta(s)} = \sum_{n=1}^{\infty} \frac{\mu(n)}{n^s} = 1 - \frac{1}{2^s} - \frac{1}{3^s} - \frac{1}{5^s} + \frac{1}{6^s} - \frac{1}{7^s} + \cdots. \qquad (2.9)$$

Using the original representation

$$\frac{1}{\zeta(s)} = \left(1 - \frac{1}{2^s}\right)\left(1 - \frac{1}{3^s}\right)\left(1 - \frac{1}{5^s}\right)\left(1 - \frac{1}{7^s}\right)\cdots, \qquad (2.10)$$

we may execute the multiplication of the various factors and so end up again with

$$1 - \frac{1}{2^s} - \frac{1}{3^s} - \frac{1}{5^s} + \frac{1}{6^s} - \frac{1}{7^s} + \frac{1}{10^s} - \cdots, \qquad (2.11)$$

which identifies the following values for μ:

$$\mu(1) = 1, \mu(2) = -1, \mu(3) = -1, \mu(4) = 0, \mu(5) = -1,$$
$$\mu(6) = 1, \mu(7) = -1, \mu(8) = 0, \mu(9) = 0, \mu(5) = 1, etc. \qquad (2.12)$$

Here is the rule:

$$\mu(n) = \begin{cases} -1 & \text{if } n \text{ contains an odd number of primes} \\ 1 & \text{if } n \text{ contains an even number of primes} \\ 0 & \text{if } n \text{ contains a quadratic prime factor} \end{cases} \qquad (2.13)$$

For example:

$$\begin{aligned} \mu(7) &= -1; 7 & \text{is a prime number} \\ \mu(66) &= -1; 66 = 2 \cdot 3 \cdot 11, & \text{odd number of primes} \\ \mu(18) &= 0; 18 = 2 \cdot 3^2, & \text{one quadratic prime number} \end{aligned} \qquad (2.14)$$

For further use we list some lower Möbius numbers:

$\mu(n) = -1$	2	3	5	7	11	13	17	19	23	29	30	31	37
$\mu(n) = 0$	4	8	9	12	16	18	20	24	25	27	28	32	36
$\mu(n)0 + 1$	1	6	10	14	15	21	22	26	33	34	35	38	39

n	1	2	3	4	5	6	7	8	9	10	11	12	13	14	15	16	17	18	19	20
$\mu(n)$	1	-1	-1	0	-1	1	-1	0	0	1	-1	0	-1	1	1	0	-1	0	-1	0

The relation between $\Pi(x)$ and $\pi(x)$ is inverted by Riemann by means of the Möbius inversion formula to obtain

$$\pi(x) = \sum_{n=1}^{\infty} \frac{\mu(n)}{n} \Pi(x^{1/n}) = \Pi(x) - \frac{1}{2}\Pi(x^{1/2}) - \frac{1}{3}\Pi(x^{1/3}) - \frac{1}{5}\Pi(x^{1/5}) + \frac{1}{6}\Pi(x^{1/6}) + \cdots .$$

(2.15)

In the final part of this section I want to discuss briefly a certain integral transform which will be of great help in the next chapter. This transformation with kernel $K(z, \xi) = \xi^{z-1}$ is known as Mellin transform, although Riemann knew about it forty years before it became known under this name.

Let us start with

$$g(z) = \int_0^{\infty} d\xi \xi^{z-1} f(\xi),$$

(2.16)

for example, with the left-hand side given by $\Gamma(s)$, $Re(s) > 0$ and $f(x) = e^{-x}$:

$$\Gamma(s) = \int_0^{\infty} dx e^{-x} x^{s-1} \text{ with inverse } e^{-x} = \frac{1}{2\pi i} \int_{a-i\infty}^{a+i\infty} ds \frac{\Gamma(s)}{x^s}.$$

(2.17)

Now we replace x by $nx (n = 1, 2, 3...)$, then multiply the equations by constants c_n and sum over n:

$$\sum_{n=1}^{\infty} \frac{c_n}{n^s} = \frac{1}{\Gamma(s)} \int_0^{\infty} x^{s-1} \left\{ \sum_{n=1}^{\infty} c_n (e^{-x})^n \right\} dx,$$

$$\sum_{n=1}^{\infty} c_n (e^{-x})^n = \frac{1}{2\pi i} \int_{a-i\infty}^{a+i\infty} \frac{\Gamma(s)}{x^s} \left\{ \sum_{n=1}^{\infty} \frac{c_n}{n^s} \right\} ds.$$

(2.18)

One can see that the Mellin transform changes the power series $\Sigma c_n(e^{-x})^n$ into a Dirichlet series $\Sigma c_n/n^s$ and the inverse of the Mellin transform changes the Dirichlet series into a power series.

In particular, if we set $c_n = 1$ for all n, then with $\Sigma(e^{-x})^n = 1/(e^x - 1)$ we obtain an integral representation of the Riemann zeta function:

$$\zeta(s) = \sum_{n=1}^{\infty} \frac{1}{n^s} = \frac{1}{\Gamma(s)} \int_0^{\infty} \frac{x^{s-1}}{e^x - 1} dx, \qquad Re(s) > 1$$

(2.19)

the inverse of which is given by

$$\frac{1}{e^x - 1} = \frac{1}{2\pi i} \int\limits_{a-i\infty}^{a+i\infty} \frac{\Gamma(s)\zeta(s)}{x^x} ds \quad (a > 1).$$ (2.20)

One of the most important formulae in Riemann's paper is given by

$$\frac{\log \zeta(s)}{s} = \int\limits_0^\infty \Pi(x) x^{-s-1} dx.$$ (2.21)

Here one recognizes for the first time the close connection between the zeta function and the function $\Pi(x)$. To understand the above formula better, let us take the logarithm of both sides of

$$\zeta(s) = \prod_p \frac{1}{1 - p^{-s}}$$ (2.22)

and using $\log(1 - x) = -x - 1/2\, x^2 - 1/3\, x^3 \cdots$ we obtain

$$\log \zeta(s) = -\sum_p \log(1 - p^{-s}) = \sum p^{-s} + \frac{1}{2}\sum p^{-2s} + \frac{1}{3}\sum p^{-3s} + \cdots.$$ (2.23)

Here we make use of the identities $(Re(s) > 1)$

$$p^{-s} = s \int\limits_p^\infty x^{-s-1} ds, \quad p^{-2s} = s \int\limits_{p^2}^\infty x^{-s-1} dx, \cdots, \quad p^{-ns} = s \int\limits_{p^n}^\infty x^{-s-1} dx, \cdots$$ (2.24)

to write

$$\log \zeta(s) = \sum_p \sum_n \frac{1}{n} p^{-ns} = \sum_p \sum_n \frac{1}{n} \cdot s \int\limits_{p^n}^\infty x^{-s-1} dx$$

$$= s \int\limits_0^\infty \Pi(x) x^{-s-1} dx.$$ (2.25)

To explain the last line, let us write

$$s \int\limits_0^\infty \Pi(x) x^{-s-1} dx = s \left\{ \left[\Pi(x)(-1)\frac{1}{s} x^{-s} \right]_0^\infty - \int\limits_0^\infty dx d\Pi \frac{x^{-s}}{-s} \right\}$$

$$= \int\limits_0^\infty x^{-s} d\Pi(x) \quad \text{(Stieltjes integral)},$$ (2.26)

where the measure $d\Pi$ has been written as the density times dx; more precisely:

$$d\Pi = \left(\frac{d\Pi}{dx}\right) dx \,, \tag{2.27}$$

where $d\Pi/dx$ is the density of primes plus 1/2-density of prime squares, plus 1/3-density of prime cubes, etc.

Let us not forget that the calculus version of the "golden formula"

$$\frac{\log \zeta(s)}{s} = \int_0^\infty \Pi(x) x^{-s-1} dx \tag{2.28}$$

has its origin in the Euler-Riemann prime product formula for the zeta function and the intelligent invention of the step function $\Pi(x)$. This name is justified because when x is the exact square of a prime, e.g., $x = 9 = 3^2$, $\Pi(x)$ jumps up one-half, since $\pi(\sqrt{x}) = \pi(3)$ jumps up 1, and so on. Note that the actual point where the jump occurs, the value of the function is halfway up the jump.

So we have derived the marvelous formula given above, which will lead us directly to the central result of Riemann's paper. But what is the inverted expression, i.e., how can we express $\Pi(x)$ in terms of $\zeta(x)$? This will be discussed in the next chapter.

Chapter 3
Riemann as an Expert in Fourier Transforms

Earlier we introduced the pair of equations

$$\frac{\log \zeta(s)}{s} = \int_0^\infty \Pi(x) x^{-s-1} dx \quad (Re(s) > 1),$$

$$\text{and} \quad \Pi(x) = \frac{1}{2\pi i} \int_{a-i\infty}^{a+i\infty} \log \zeta(s) x^s \frac{ds}{s} \quad (a > 1), \tag{3.1}$$

when we discussed the Mellin transform. Let us see how Riemann reached the same result much earlier by employing the Fourier inversion formula:

$$\varphi(x) = \frac{1}{2\pi} \int_{-\infty}^{+\infty} \left[\int_{-\infty}^{+\infty} \varphi(\lambda) e^{i(x-\lambda)\mu} d\lambda \right] d\mu. \tag{3.2}$$

When we write

$$\varphi(x) = \int_{-\infty}^{+\infty} \phi(\mu) e^{i\mu x} d\mu, \tag{3.3}$$

we can consider $\phi(\mu)$ as coefficients of an expansion defined by

$$\phi(\mu) = \frac{1}{2\pi} \int_{-\infty}^{+\infty} \varphi(\lambda) e^{-i\lambda\mu} d\lambda. \tag{3.4}$$

Now let $s = a + i\mu$, $a = const. > 1$ and μ be a real variable.

© The Author(s), under exclusive license to Springer Nature Switzerland AG 2021
W. Dittrich, *Reassessing Riemann's Paper*,
SpringerBriefs in History of Science and Technology,
https://doi.org/10.1007/978-3-030-61049-4_3

Then with $\lambda = \log x$ and $\varphi(x) = 2\Pi(e^x)e^{-ax}$, we obtain

$$
\begin{array}{cc}
x = e^\lambda \\
\frac{dx}{x} = d\lambda
\end{array}
:\quad
\frac{\log \zeta(a+i\mu)}{a+i\mu} = \int_{-\infty}^{+\infty} \Pi(e^\lambda)e^{-(a+i\mu)\lambda}d\lambda
$$

$$
=: \phi(\mu) = \frac{1}{2\pi}\int_{-\infty}^{+\infty} \varphi(\lambda)e^{-i\mu\lambda}d\lambda. \tag{3.5}
$$

Hence we can continue to write

$$
(\varphi(x)) = 2\pi\Pi(e^x)e^{-ax} = \int_{-\infty}^{+\infty} \frac{\log \zeta(a+i\mu)}{a+i\mu}e^{i\mu x}d\mu \tag{3.6}
$$

and using $e^x = y$, then $y \to x, s = a+i\mu, ds = id\mu, d\mu = 1/i \cdot ds$ we finally obtain

$$
\Pi(x) = \frac{1}{2\pi i}\int_{a-i\infty}^{a+i\infty} \log \zeta(s)x^s\frac{ds}{s} \quad (a > 1), \tag{3.7}
$$

which is the desired result.

From here on we can directly arrive at Riemann's main result of his 1859 paper. However, for the time being we have to accept two of Riemann's novel quantities (details will be reported later): The entire function $\xi(s)$ ($\zeta(s)$ is not an entire function) and the product formula for the ξ function:

$$
\xi(s) = \frac{1}{2}s(s-1)\pi^{-\frac{s}{2}}\Gamma\left(\frac{s}{2}\right)\zeta(s), \qquad \Gamma\left(\frac{s}{2}\right) = \frac{2}{s}\Gamma\left(1+\frac{s}{2}\right)
$$

$$
= (s-1)\pi^{-\frac{s}{2}}\Gamma\left(1+\frac{s}{2}\right)\zeta(s) \tag{3.8}
$$

and

$$
\xi(s) = \frac{1}{2}\prod_{\rho}\left(1-\frac{s}{\rho}\right), \tag{3.9}
$$

with ρ the zeros of the zeta function (equal to the zeros of ξ).

So, taking the logarithm of both sides, we obtain

$$
-\log 2 + \sum_{\rho}\log\left(1-\frac{s}{\rho}\right) = \log(s-1) - \frac{s}{2}\log\pi + \log\Gamma\left(1+\frac{s}{2}\right) + \log\zeta(s)
$$

$$
\text{or}\quad \log\zeta(s) = \sum_{\rho}\log\left(1-\frac{s}{\rho}\right) - \log 2 - \log\Gamma\left(1+\frac{s}{2}\right) + \frac{s}{2}\log\pi - \log(s-1).
$$

$$
\tag{3.10}
$$

The first term on the right-hand side gives us the searched-for connection of the non-trivial zeta zeros with $\Pi(x)$. This becomes evident when we write

$$\Pi(x) = \frac{1}{2\pi i} \int\limits_{a-i\infty}^{a+i\infty} \frac{\log \zeta(s)}{s} x^s ds \qquad (3.11)$$

with $\log \zeta(s)$ taken from above. Here, then, is Riemann's result:

$$\Pi(x) = Li(x) - \sum_{\rho} Li(x^\rho) + \log\left(\frac{1}{2}\right) + \int\limits_{x}^{\infty} \frac{dt}{t(t^2 - 1)\log t}, \qquad x > 1.$$

$$(3.12)$$

The sum over ρ is to be understood as

$$\sum_{\operatorname{Im}\rho>0} (Li(x^\rho) + Li(x^{1-\rho})) \qquad (3.13)$$

and $Li(x)$ denotes the logarithmic integral (see below).

This calculated expression for $\Pi(x)$ is then used in the formula

$$\pi(x) = \sum_{n=1}^{\infty} \frac{\mu(n)}{n} \Pi(x^{1/n}) = \Pi(x) - \frac{1}{2}\Pi(x^{1/2}) - \frac{1}{3}\Pi(x^{1/3}) - \frac{1}{5}\Pi(x^{1/5}) + \frac{1}{6}\Pi(x^{1/6}) + \cdots .$$

$$(3.14)$$

This is Riemann's great achievement, the explicit, exact calculation of the prime number counting function $\pi(x)$.

Let us rewrite Riemann's result more explicitly:

$$\Pi(x) = Li(x) - \sum_{\operatorname{Im}\rho>0} (Li(x^\rho) + Li(x^{1-\rho})) - \log 2 + \int\limits_{x}^{\infty} \frac{dt}{t(t^2 - 1)\log t}, \qquad x > 1$$

$$(3.15)$$

with

$$Li(x) = \lim_{\epsilon \to 0} \left[\int\limits_{0}^{1-\epsilon} \frac{dt}{\log t} + \int\limits_{1+\epsilon}^{x} \frac{dt}{\log t} \right]. \qquad (3.16)$$

If we differentiate $\Pi(x)$ we obtain

$$d\Pi = \left[\frac{1}{\log x} - \sum_{\operatorname{Re}\alpha>0} \frac{2\cos(\alpha \log x)}{x^{1/2}\log x} - \frac{1}{x(x^2 - 1)\log x} \right] dx \qquad x > 1. \quad (3.17)$$

α ranges over all values such that $\rho = 1/2 + i\alpha$; in other words, $\alpha = -i(\rho - 1/2)$ where ρ ranges over all roots, so that

$$x^{\rho-1} + x^{-\rho} = x^{-\frac{1}{2}}\left[x^{i\alpha} + x^{-i\alpha}\right] = 2x^{-\frac{1}{2}}\cos(\alpha \log x). \qquad (3.18)$$

The Riemann hypothesis says that the α's are all real.

Again, by the definition of Π, the measure $d\Pi$ is dx times the density of primes plus 1/2 the density of prime squares, plus 1/3 the density of prime cubes plus, etc. Thus $1/(\log x)$ alone should not be considered an approximation only to the density of primes as Gauß suggested, but rather to $d\Pi/dx$, i.e., to the density of primes plus 1/2 the density of prime squares, plus, etc.

A fairly good approximation neglects the last term in $d\Pi$. It is the number of α's which is significant in $d\Pi$ which Riemann meant to study empirically to see the influence of the "periodic terms" on the distribution of primes. With the above equations we have reached the end of Riemann's famous paper of 1859.

We have, however, left out a number of revolutionary results to which we want to turn to now.

Chapter 4
On the Way to Riemann's Entire Function $\xi(s)$

Let us begin with the integral representation of Euler's Γ function:

$$\Gamma(s) = \int_0^\infty x^{s-1} e^{-x} dx,$$

$$s \to \frac{s}{2}: \quad \Gamma\left(\frac{s}{2}\right) = \int_0^\infty x^{\frac{s}{2}-1} e^{-x} dx,$$

$$x = \pi t n^2: \quad \Gamma\left(\frac{s}{2}\right) = \int_0^\infty (\pi t n^2)^{\frac{s}{2}-1} e^{-\pi t n^2} \pi n^2 dt,$$

$$\Gamma\left(\frac{s}{2}\right) \pi^{-\frac{s}{2}} \frac{1}{n^s} = \int_0^\infty e^{-\pi t n^2} t^{\frac{s}{2}} \frac{dt}{t},$$

$$\text{Take} \sum_{n=1}^\infty: \quad \Gamma\left(\frac{s}{2}\right) \pi^{-\frac{s}{2}} \zeta(s) = \int_0^\infty \psi(t) t^{\frac{s}{2}} \frac{dt}{t}, \qquad Re(s) > 1,$$

$$\psi(t) = \sum_{n=1}^\infty e^{-\pi t n^2}. \tag{4.1}$$

The last equation defines one of Jacobi's ϑ functions:

$$\Theta(x) := \vartheta_3(0, ix) = \sum_{n=-\infty}^{+\infty} e^{-\pi x n^2}, \qquad \psi(x) = \sum_{n=1}^\infty e^{-\pi x n^2}, \qquad \Theta(x) = 2\psi(x) + 1. \tag{4.2}$$

W. Dittrich, *Reassessing Riemann's Paper*,
SpringerBriefs in History of Science and Technology,
https://doi.org/10.1007/978-3-030-61049-4_4

Also let me quote without proof the Jacobi identity:

$$\Theta(x) = \frac{1}{\sqrt{x}} \Theta\left(\frac{1}{x}\right), \qquad x > 0. \tag{4.3}$$

One can then easily verify that

$$\frac{1 + 2\psi(x)}{1 + 2\psi\left(\frac{1}{x}\right)} = \frac{1}{\sqrt{x}}, \tag{4.4}$$

so that

$$\psi\left(\frac{1}{x}\right) = \frac{1}{2}\Theta\left(\frac{1}{x}\right) - \frac{1}{2} = \frac{1}{\sqrt{2}}\sqrt{x}\Theta(x) - \frac{1}{2} = \sqrt{x}\psi(x) + \frac{\sqrt{x}}{2} - \frac{1}{2}. \tag{4.5}$$

Now we are going to calculate the following integral, which will give us one of Riemann's wonderful results.

Using $\Psi(x) = x^{-1/2}\Psi(1/x) - 1/2 + 1/2x^{-1/2}$ and splitting the integral apart at 1, we obtain

$$\int_0^\infty \Psi(x) x^{s/2} \frac{dx}{x} = \int_1^\infty \Psi(x) x^{s/2} \frac{dx}{x} + \int_0^1 \Psi\left(\frac{1}{x}\right) x^{\frac{s-1}{2}} \frac{dx}{x} + \frac{1}{2}\int_0^1 \left(x^{\frac{s-1}{2}} - x^{\frac{s}{2}}\right) \frac{dx}{x}. \tag{4.6}$$

In the last two integrals we substitute $x \rightarrow 1/x$ and so we get

$$\int_0^\infty \Psi(x) x^{\frac{s}{2}} \frac{dx}{x} = \int_1^\infty \Psi(x) \left[x^{\frac{s}{2}} + x^{\frac{1}{2}(1-s)}\right] \frac{dx}{x} + \frac{1}{2}\int_1^\infty \left[x^{\frac{1}{2}(1-s)} - x^{-\frac{s}{2}}\right] \frac{dx}{x}$$

$$\int_1^\infty dx \left[x^{-\frac{s}{2}-\frac{1}{2}}\right] = -\frac{2}{s-1},$$

$$\int_1^\infty dx \left[x^{-\frac{s}{2}-1}\right] = \frac{2}{s},$$

$$= \int_1^\infty \Psi(x) \left(x^{\frac{s}{2}-1} + x^{-\frac{s}{2}-\frac{1}{2}}\right) dx = \frac{1}{s} + \frac{1}{s-1}. \tag{4.7}$$

Here, then, is the important formula contained in Riemann's paper:

$$\Gamma\left(\frac{s}{2}\right)\pi^{-\frac{s}{2}}\zeta(s) = \int_1^\infty \Psi(x)\left(x^{\frac{s}{2}-1} + x^{-\frac{s}{2}-\frac{1}{2}}\right)dx - \frac{1}{s(1-s)}.$$

$$\begin{array}{r} pole\,\Gamma : s = 0 \\ pole\,\zeta : s = 1 \end{array} \qquad (4.8)$$

Notice that there is no change of the right-hand side under $s \to 1 - s$! $\pi^{-s/2}$ $\Gamma(s/2)\zeta(s)$ has simple poles at $s = 0$ and $s = 1$. To remove these poles, we multiply by $1/2s(s-1)$. This is the reason why Riemann defines

$$\xi(s) = \frac{1}{2}s(s-1)\pi^{-\frac{s}{2}}\Gamma\left(\frac{s}{2}\right)\zeta(s), \qquad (4.9)$$

which is an entire function ($\zeta(s)$ is a meromorphic function.) Obviously we have $\xi(s) = \xi(1-s)$ and the functional equation

$$\Gamma\left(\frac{s}{2}\right)\pi^{-\frac{s}{2}}\zeta(s) = \Gamma\left(\frac{1-s}{2}\right)\pi^{-\frac{1}{2}(1-s)}\zeta(1-s). \qquad (4.10)$$

We obtain the right-hand side by the left-hand side by replacing s by $(1-s)$.

Now we can continue to write for $\xi(s)$

$$\xi(s) = \frac{1}{2} - \frac{s(1-s)}{2}\int_1^\infty \Psi(x)\left(x^{\frac{s}{2}} + x^{\frac{1}{2}(1-s)}\right)\frac{dx}{x}$$

$$= \frac{1}{2} - \frac{s(1-s)}{2}\int_1^\infty \frac{d}{dx}\left\{\Psi(x)\left[\frac{x^{\frac{s}{2}}}{\frac{s}{2}} + \frac{x^{\frac{1}{2}(1-s)}}{\frac{1}{2}(1-s)}\right]\right\}dx$$

$$+ \frac{s(1-s)}{2}\int_1^\infty \Psi'(x)\left[\frac{x^{\frac{s}{2}}}{\frac{s}{2}} + \frac{x^{\frac{1}{2}(1-s)}}{\frac{1}{2}(1-s)}\right]ds$$

$$= \frac{1}{2} + \frac{s(1-s)}{2}\Psi(1)\left[\frac{2}{3} + \frac{2}{1-s}\right]$$

$$+ \int_1^\infty \Psi'(x)\left[(1-s)x^{\frac{s}{2}} + sx^{\frac{1}{2}(1-s)}\right]dx$$

$$= \frac{1}{2} + \Psi(1) + \int_1^\infty x^{\frac{s}{2}}\Psi'(x)\left[(1-s)x^{\frac{1}{2}(s-1)-1} + sx^{-\frac{s}{2}-1}\right]dx$$

$$= \frac{1}{2} + \Psi(1) + \int_1^\infty \frac{d}{dx}\left[x^{\frac{3}{2}}\Psi'(x)\left(-2x^{\frac{1}{2}(s-1)} - 2x^{-\frac{s}{2}}\right)\right]dx$$

$$-\int_1^\infty \frac{d}{dx}\left[x^{\frac{3}{2}}\Psi'(x)\right]\left[-2x^{\frac{1}{2}(s-1)} - 2x^{-\frac{s}{2}}\right]dx$$

$$= \frac{1}{2} + \Psi(1) - \Psi'(1)[-2-2] + \int_1^\infty \frac{d}{dx}\left[x^{\frac{3}{2}}\Psi'(x)\right]\left(2x^{\frac{1}{2}(s-1)} + 2x^{-\frac{s}{2}}\right)dx.$$

$$(4.11)$$

Differentiation of

$$2\Psi(x) + 1 = x^{-\frac{1}{2}}\left[2\Psi\left(\frac{1}{x}\right) + 1\right] \tag{4.12}$$

easily gives

$$\frac{1}{2} + \Psi(1) + 4\Psi'(1) = 0 \tag{4.13}$$

and using this puts the formula in the final form:

$$\xi(s) = 4\int_1^\infty \frac{d}{dx}\left[x^{\frac{3}{2}}\Psi'(x)\right]x^{-\frac{1}{4}}\cosh\left[\frac{1}{2}\left(s - \frac{1}{2}\right)\log x\right]dx, \tag{4.14}$$

or, as Riemann writes it ($s = 1/2 + it$; $1/2$ is Riemann's conjecture!):

$$\Xi(t) = \xi\left(\frac{1}{2} + it\right) = 4\int_1^\infty \frac{d}{dx}\left[x^{\frac{3}{2}}\psi'(x)\right]x^{-\frac{1}{4}}\cos\left(\frac{t}{2}\log x\right)dx. \tag{4.15}$$

With

$$\frac{d}{dx}\left[x^{3/2}\psi'(x)\right] = \sum_{n=1}^\infty\left(n^4\pi^2 x - \frac{3}{2}n^2\pi\right)x^{1/2}\exp(-n^2\pi x) \tag{4.16}$$

and

$$v = \frac{1}{2}\log x \tag{4.17}$$

and then $v = 2u$, we can also write $\Xi\left(\frac{t}{2}\right)$ as a Fourier transform

$$\Xi\left(\frac{t}{2}\right) = 8\int_0^\infty du\, \Phi(u)\cos(ut) \tag{4.18}$$

with

$$\Phi(u) = \sum_{n=1}^{\infty} \pi n^2 \left(2n^2\pi \exp(4u) - 3\right) \exp(5u - n^2\pi \exp(4u)). \qquad (4.19)$$

If $\cosh[1/2(s-1/2)\log x]$ is expanded in the usual power series

$$\cos hy = \frac{1}{2}\left(e^y + e^{-y}\right) = \sum \frac{y^{2n}}{(2n)!}, \qquad (4.20)$$

we can write

$$\xi(s) = \sum_{n=0}^{\infty} a_{2n}\left(s - \frac{1}{2}\right)^{2n}, \qquad (4.21)$$

where

$$a_{2n} = 4\int_{1}^{\infty} \frac{d}{dx}\left[x^{3/2}\Psi'(x)\right]x^{-\frac{1}{4}}\frac{\left(\frac{1}{2}\log x\right)^{2n}}{(2n)!}dx. \qquad (4.22)$$

Let us return to

$$\xi(s) = \frac{1}{2}s(s-1)\pi^{-\frac{s}{2}}\Gamma\left(\frac{s}{2}\right)\zeta(s), \qquad (4.23)$$

with

$$\pi^{-\frac{s}{2}}\Gamma\left(\frac{s}{2}\right)\zeta(s) = \frac{1}{s(s-1)} + \int_{1}^{\infty} \Psi(x)\left(x^{\frac{s}{2}-1} + x^{-\frac{s}{2}-\frac{1}{2}}\right)dx, \qquad (4.24)$$

and write the right-hand side in terms of $s = 1/2 + it$, which makes use of Riemann's conjecture $Re(s) = 1/2$. Since the details of the substitution are trivial, we merely give the final result:

$$\xi\left(\frac{1}{2} + it\right) = \frac{1}{2}\left(\frac{1}{2} + it\right)\left(it - \frac{1}{2}\right)\pi^{-\frac{1}{4} - i\frac{t}{2}}\Gamma\left(\frac{1}{4} + i\frac{t}{2}\right)\zeta\left(\frac{1}{2} + it\right)$$

$$= \frac{-\left(t^2 + \frac{1}{4}\right)}{\left[2(\sqrt{\pi})^{\frac{1}{2}+it}\right]}\Gamma\left(\frac{1}{4} + \frac{it}{2}\right)\zeta\left(\frac{1}{2} + it\right). \qquad (4.25)$$

In particular,

$$\xi\left(\frac{1}{2}\right) = \frac{-1}{(8\pi^{1/4})}\Gamma\left(\frac{1}{4}\right)\zeta\left(\frac{1}{2}\right) \qquad (4.26)$$

with

$$\zeta\left(\frac{1}{2}\right) = -1.4603545088, \quad \Gamma\left(\frac{1}{4}\right) = \sqrt{2\varpi 2\pi} = 3.6256099082, \qquad (4.27)$$

where Gauss' lemniscate constant is given by

$$\varpi = 2.62205755429. \tag{4.28}$$

Altogether:

$$\xi\left(\frac{1}{2}\right) = 0.4971207781 = a_0, \tag{4.29}$$

which is the minimum for the real valued $\xi(s)$ at $s = 1/2$. By the way $\xi(0) = \xi(1) = -\zeta(0) = 1/2$. The above result can also be written as

$$\Xi(t) := \xi\left(\frac{1}{2} + it\right) = \frac{1}{2} - \left(t^2 + \frac{1}{4}\right) \int_1^\infty \Psi(x) x^{-\frac{3}{4}} \cos\left(\frac{t}{2}\log x\right) dx. \tag{4.30}$$

The right-hand side of this equation tells us that because $t \in R_e$, $x \in R_e$ and $\log x \in R_e$, we have

$$\mathrm{Im}\xi\left(\frac{1}{2} + it\right) = 0, \quad i.e., \quad \xi\left(\frac{1}{2} + it\right) \equiv \Xi(t) \in R_e. \tag{4.31}$$

Since $\Xi(t) = \xi(1/2 + it)$ for $t \to \infty$ changes its sign infinitely often, $\xi(s)$ (and $\zeta(s)$) must have infinitely many zeros on $Re(s) = 1/2$.

There is another useful form $\xi(s)$ that starts with its original definition:

$$\xi(s) = \frac{s(s-1)}{2} \Gamma\left(\frac{s}{2}\right) \pi^{-\frac{s}{2}} \zeta(s)$$

$$= e^{\ln \Gamma(\frac{s}{2})} \pi^{-\frac{s}{2}} \frac{s(s-1)}{2} \zeta(s). \tag{4.32}$$

Then, setting $s = 1/2 + it$, we have

$$\xi\left(\frac{1}{2} + it\right) = e^{\ln \Gamma\frac{(\frac{1}{2}+it)}{2}} \pi^{-\frac{\frac{1}{2}+it}{2}} \frac{1}{2}\left(\frac{1}{2} + it\right)\left(\frac{1}{2} + it - 1\right)\zeta\left(\frac{1}{2} + it\right)$$

$$= \left[e^{R_e \ln \frac{(\frac{1}{2}+it)}{2}} \pi^{-\frac{1}{4}} \cdot \frac{-t^2 - \frac{1}{4}}{2}\right] \left[e^{i \mathrm{Im} \ln \Gamma \frac{(\frac{1}{2}+it)}{2}} \pi^{-\frac{it}{2}} \zeta\left(\frac{1}{2} + it\right)\right]$$

$$= \left[-e^{R_e \ln \Gamma\left(\frac{\frac{1}{2}+it}{2}\right)} \pi^{-\frac{1}{4}} \frac{t^2 + \frac{1}{4}}{2}\right] \left[e^{i \mathrm{Im} \ln \Gamma\left(\frac{\frac{1}{2}+it}{2}\right)} \pi^{-\frac{it}{2}} \zeta\left(\frac{1}{2} + it\right)\right]. \tag{4.33}$$

Notice that the first factor in the square brackets is negative. For the second factor we have

$$Z(t) = e^{i\vartheta(t)}\zeta\left(\frac{1}{2}+it\right), \qquad \vartheta(t) = \operatorname{Im}\ln\Gamma\left(\frac{\frac{1}{2}+it}{2}\right) - \frac{t}{2}\ln\pi. \qquad (4.34)$$

Thus, $Z(t)$ has always the opposite sign compared to the ξ function.

Now we have to compute $\vartheta(t)$ and $\zeta(1/2+it)$. For numerical analysis it is sufficient to use

$$\vartheta(t) \sim \frac{t}{2}\log\frac{t}{2\pi} - \frac{t}{2} - \frac{\pi}{8} + \frac{1}{48t}, \qquad (4.35)$$

which one can then apply to compute the roots of $\xi(s)$ on the critical line.

Chapter 5
The Product Representation of $\xi(s)$ and $\zeta(s)$ by Riemann (1859) and Hadamard (1893)

Riemann's goal (before Weierstrass!) was to prove that $\xi(s)$ can be expanded as an infinite product

$$\xi(s) = \xi(0) \prod_{\rho} \left(1 - \frac{s}{\rho}\right), \tag{5.1}$$

where ρ ranges over all the roots of $\xi(\rho) = 0$. He did not really prove this formula, but he was right, as shown much later by Hadamard. But one has to admit that Riemann must have had a strong inkling of the product formula Weierstrass was soon to introduce as an essential representation into the function theory, here the entire functions, i.e., functions that can be determined by their zeros.

As a brief reminder, here is Weierstrass' product representation of the Γ function:

$$\Gamma(x) = e^{-\gamma x} \frac{1}{x} \prod_{k=1}^{\infty} \frac{e^{\frac{x}{k}}}{\left(1 + \frac{x}{k}\right)}, \tag{5.2}$$

where γ is the Euler-Mascheroni constant,

$$\gamma = \lim_{n \to \infty} \left[\sum_{k=1}^{n} \frac{1}{k} - \log n \right] \simeq 0.5772157. \tag{5.3}$$

From this product formula follows, with the aid of

$$\Gamma(x)\Gamma(1-x) = \Gamma(x)(-x)\Gamma(-x) = \frac{\pi}{\sin(\pi x)}, \tag{5.4}$$

© The Author(s), under exclusive license to Springer Nature Switzerland AG 2021
W. Dittrich, *Reassessing Riemann's Paper*,
SpringerBriefs in History of Science and Technology,
https://doi.org/10.1007/978-3-030-61049-4_5

the product representation of $\sin(\pi x)$:

$$\sin(\pi x) = -\frac{\pi}{x}\frac{1}{\Gamma(x)\Gamma(-x)} = -\frac{\pi}{x}\left(e^{\gamma x}x\prod_{k=1}^{\infty}\frac{(1+\frac{x}{k})}{e^{\frac{x}{n}}}\right)\left(e^{-\gamma x}(-x)\prod_{k>1}^{\infty}\frac{(1-\frac{x}{k})}{e^{-\frac{x}{k}}}\right)$$

$$= \pi x\prod_{k=1}^{\infty}\left(1-\frac{x^2}{k^2}\right), \tag{5.5}$$

a polynomial of infinite degree. Similarly, Euler thought of $\sin(\pi x)$ as a polynomial of infinite degree when he conjectured, and finally proved, the formula for $\sin(\pi x)$.

So, why not think of $\xi(s)$ as a polynomial of infinite degree and write down a product formula determined by its infinite zeros ρ? This is what Hadamard had done in 1893 in a paper in which he studied entire functions and their representations as infinite products—like Weierstrass. He was able to prove that Riemann's product formula was correct:

$$\xi(s) = \xi(0)\prod_{\rho}\left(1-\frac{s}{\rho}\right). \tag{5.6}$$

$\xi(s)$ is an entire function. The infinite product is understood to be taken in an order which pairs each root ρ with the corresponding root $1-\rho$. Hadamard's proof of the product formula for ξ was called by von Mangoldt "the first real progress in the field in 34 years," that is, the first since Riemann.

Hadamard showed that it is possible to construct the ζ function as an infinite product, given its zeros:

$$\zeta(s) = f(s)\prod_{\zeta(\rho)=0}\left(1-\frac{s}{\rho}\right)e^{\frac{s}{\rho}}, \qquad f(s) = \frac{1}{2(s-1)}\left(\frac{2\pi}{e}\right)^s. \tag{5.7}$$

Hence, including the trivial as well as the non-trivial zeros he obtains

$$\zeta(s) = \frac{1}{2(s-1)}\left(\frac{2\pi}{e}\right)^s\prod_{n=1}^{\infty}\left(1+\frac{s}{2n}\right)e^{-\frac{s}{2n}}\cdot\prod_{\rho}\left(1-\frac{s}{\rho}\right)e^{\frac{s}{\rho}}. \tag{5.8}$$

For the first product we use the product representation given by Weierstrass:

$$\frac{1}{\Gamma(1+s)} = e^{\gamma s}\prod_{n=1}^{\infty}\left(1+\frac{s}{n}\right)e^{-\frac{s}{n}}, \tag{5.9}$$

and so obtain the Hadamard product formula, which is convergent in $C \setminus \{1\}$:

$$\zeta(s) = \frac{e^{(\log 2\pi - 1 - \frac{\gamma}{2})s}}{2(s-1)\Gamma\left(1+\frac{s}{2}\right)}\prod_{\rho}\left(1-\frac{s}{\rho}\right)e^{\frac{s}{\rho}}. \tag{5.10}$$

A slightly simplified form of the Hadamard product is

$$\zeta(s) = \frac{\pi^{s/2}}{2(s-1)\Gamma\left(1+\frac{s}{2}\right)} \prod_\rho \left(1 - \frac{s}{\rho}\right).$$ (5.11)

Here we took pairs of roots ρ and $-\rho$ together so the exponents $e^{-s/\rho}$ cancel.

The last expression shows the the ζ function can be completely constructed by its roots (Riemann's specialty) and the singularity at $s = 1$. However, to obtain absolute convergence, we have to introduce ρ and $-\rho$ pairwise in the product.

Now, we remember Riemann's entire function $\xi(s)$ and how it is related to the (non-entire) ζ function:

$$\xi(s) = \frac{s(s-1)}{2} \Gamma\left(\frac{s}{2}\right) \pi^{-\frac{s}{2}} \zeta(s).$$ (5.12)

Then

$$\frac{s(s-1)}{2} \pi^{-\frac{s}{2}} \Gamma\left(\frac{s}{2}\right) \cdot \frac{\pi^{s/2}}{2(s-1)\Gamma\left(1+\frac{s}{2}\right)} \prod_\rho \left(1 - \frac{s}{\rho}\right), \qquad \Gamma\left(1+\frac{s}{2}\right) = \frac{s}{2}\Gamma\left(\frac{s}{2}\right)$$ (5.13)

or

$$\xi(s) = \frac{1}{2} \prod_\rho \left(1 - \frac{s}{\rho}\right)$$ (5.14)

and using $\xi(0) = \frac{1}{2}$, we have indeed

$$\xi(s) = \xi(0) \prod_\rho \left(1 - \frac{s}{\rho}\right),$$ (5.15)

which is Riemann's result of 1859!

Since the zeros of $\zeta(s)$ and $\xi(s)$ in the critical strip are identical, we can also write

$$\zeta(s) = \frac{\pi^{s/2}}{2(s-1)\Gamma\left(1+\frac{s}{2}\right)} \prod_\rho \left(1 - \frac{s}{\rho}\right)\left(1 - \frac{s}{1-\rho}\right)$$

$$= \frac{\pi^{s/2}}{2(s-1)\Gamma\left(1+\frac{s}{2}\right)} \left(1 - \frac{s}{\frac{1}{2}+14.134i}\right)\left(1 - \frac{s}{\frac{1}{2}-14.134i}\right)\left(1 - \frac{s}{\frac{1}{2}+21.022i}\right)(\cdots),$$ (5.16)

where we have used the first zeros on the $Re(s) = 1/2$ axis.

Chapter 6
Derivation of von Mangoldt's Formula for $\Psi(x)$

Hans von Mangoldt (1854 – 1925)

© The Author(s), under exclusive license to Springer Nature Switzerland AG 2021
W. Dittrich, *Reassessing Riemann's Paper*,
SpringerBriefs in History of Science and Technology,
https://doi.org/10.1007/978-3-030-61049-4_6

There is another, more modern version of an equivalent to Riemann's formula for $\Pi(x)$, i.e.,

$$\Pi(x) = Li(x) - \sum_{\rho} Li(x^{\rho}) + \log \xi(0) + \int\limits_{x}^{\infty} \frac{dt}{t(t^2-1)\log t} \qquad (x > 1). \qquad (6.1)$$

This is von Mangoldt's formula for $\Psi(x)$, which contains essentially the same information as Riemann's $\Pi(x)$. On the way to the explicit formula for $\Psi(x)$, we need a special representation of the discontinuity function. So let us begin very simply by verifying

$$\frac{1}{s-\beta} = \int\limits_{1}^{\infty} x^{-s} x^{\beta-1} dx, \qquad Re(s-\beta) > 0,$$

$$x = e^{\lambda} := \int\limits_{0}^{\infty} e^{-\lambda s} e^{\lambda(\beta-1)} e^{\lambda} d\lambda = \int\limits_{0}^{\infty} e^{-\lambda s} e^{\lambda \beta} d\lambda,$$

$$s = a + i\mu = \int\limits_{0}^{\infty} e^{-\lambda(a+i\mu)} e^{\lambda \beta} \alpha \lambda,$$

$$\frac{1}{a+i\mu-\beta} = \int\limits_{0}^{\infty} e^{-i\lambda\mu} e^{\lambda(\beta-a)} d\lambda, \qquad a > Re\beta,$$

$$\int\limits_{-\infty}^{+\infty} \frac{1}{a+i\mu-\beta} e^{i\mu x} d\mu = \int\limits_{-\infty}^{+\infty} e^{i\mu x} d\mu \int\limits_{0}^{\infty} e^{-i\lambda\mu} e^{\lambda(\beta-a)} d\lambda$$

$$= \int\limits_{-\infty}^{+\infty} \left[\int\limits_{0}^{\infty} e^{i(x-\lambda)\mu} d\mu \right] e^{\lambda(\beta-a)} d\lambda$$

$$= \int\limits_{-\infty}^{+\infty} 2\pi\delta(x-\lambda) e^{\lambda(\beta-a)} d\lambda$$

$$= \begin{cases} 2\pi e^{x(\beta-a)} , & x > 0 \\ 0 , & x < 0 \end{cases} \qquad (6.2)$$

So far we have

$$\frac{1}{2\pi} \int\limits_{-\infty}^{+\infty} \frac{1}{a+i\mu-\beta} e^{x(a+i\mu)} d\mu = \begin{cases} e^{x\beta} , & x > 0 \\ 0 , & x < 0 \end{cases} \qquad (6.3)$$

With $e^x = y$ and $s = a + i\mu$, we obtain the discontinuity factor (step function)

$$\frac{1}{2\pi i} \int_{a-i\infty}^{a+i\infty} \frac{1}{s-\beta} y^s ds = \begin{cases} y^\beta & , y > 1 \\ 0 & , y < 1 \end{cases} \overset{\beta=0}{=} \begin{cases} 1 & , y > 1 \\ \frac{1}{2} & , y = 0 \\ 0 & , y < 1 \end{cases} \quad a > 0 . \tag{6.4}$$

Now we go back to the Euler-Riemann zeta function,

$$\zeta(z) = \prod_{p \in P} \frac{1}{1-p^{-z}}, \qquad Re(z) > 1 \tag{6.5}$$

and take the logarithm:

$$\log \zeta(z) = -\sum_p \log(1-p^{-z}) = -\sum_p \log \left(1 - e^{-z\log p}\right),$$

$$\frac{d}{dz} \log \zeta(z) = -\sum_p \frac{1}{1-p^{-z}} \frac{d}{dz} \left(1 - e^{-z\log p}\right) = -\sum_p \frac{1}{1-p^{-z}} \log p \cdot p^{-z}$$

$$= -\sum_p \frac{p^{-z}}{1-p^{-z}} \log p = -\sum_p \sum_{\nu=1}^{\infty} p^{-\nu z} \log p$$

$$= \frac{\zeta'(z)}{\zeta(z)} .$$

$$\frac{x^z}{z} : \quad \frac{x^z}{z} \sum_p \sum_{\nu=1}^{\infty} \frac{\log p}{p^{\nu z}} = \sum_p \sum_{\nu=1}^{\infty} \left(\frac{x}{p^\nu}\right)^z \frac{\log p}{z} = -\frac{\zeta'(z)}{\zeta(z)} \cdot \frac{x^z}{z} ,$$

$$\frac{1}{2\pi i} \int_{a-i\infty}^{a+i\infty} \sum_{p,\nu=1}^{\infty} \left(\frac{x}{p^\nu}\right)^z \frac{\log p}{z} = \frac{1}{2\pi i} \int_{a-i\infty}^{a+i\infty} -\frac{\zeta'(x)}{\zeta(z)} \frac{x^z}{z} dz$$

$$\text{or} \quad \sum_p^{\infty} \log p \frac{1}{2\pi i} \int_{a-i\infty}^{a+i\infty} \left(\frac{x}{p^\nu}\right)^z \frac{1}{z} dz = \frac{1}{2\pi i} \int_{a-i\infty}^{a+i\infty} -\frac{\zeta'(z)}{\zeta(z)} \frac{x^z}{z} dz$$

$$y = \frac{x}{p^\nu} : \quad \sum_p^{\infty} \log p \frac{1}{2\pi i} \int_{a-i\infty}^{a+i\infty} \frac{y^z}{z} dz = \frac{1}{2\pi i} \int_{a-i\infty}^{a+i\infty} -\frac{\zeta'(z)}{\zeta(z)} \frac{x^z}{z} dz . \tag{6.6}$$

Here we use the 1 of the discontinuity factor on the left-hand side and so obtain the Chebyshev function $\Psi(x)$:

$$\Psi(x) = \sum_{p^\nu < x} \log p = \frac{1}{2\pi i} \int_{a-i\infty}^{a+i\infty} -\frac{\zeta'(z)}{\zeta(z)} \frac{x^z}{z} dz . \tag{6.7}$$

So one has to sum the logarithm of all primes up to x. $p^\nu > x$ would mean $y < 1$, but for this case the discontinuity formula gives zero.

The integral of the right-hand side can be evaluated with the aid of the theorem of residues. The contributions to the residues of $\zeta'(z)/\zeta(z) \cdot x^z/z$ come from

Singularity	Reason	Residue
0	$\frac{x^z}{z}$	$\frac{\zeta'(0)}{\zeta(0)} = \frac{-\frac{1}{2}\log 2\pi}{-\frac{1}{2}} = \log(2\pi)$
1	pole of ζ $\frac{\zeta'(z)}{\zeta(z)} = -\frac{1}{z-1} + \gamma + \cdots$	$\lim_{z \to 1}(z-1)\left(\frac{-1}{z-1} + \mathcal{O}(1)\right)\frac{x^z}{z} = \frac{-x^1}{1} = -x$
$-2, -4, -6, \cdots$	trivial zeros of $\zeta(z)$	$\frac{1}{2}x^{-2}, \frac{1}{4}x^{-4}, \frac{1}{6}x^{-6}, \cdots$ $\sum_{n=1}^{\infty} \frac{x^{-2n}}{2n} = \frac{1}{2}\log\left(1 - \frac{1}{x^2}\right)$
ρ	nontrivial zeros of $\zeta(z)$	$\frac{x^\rho}{\rho}$

$$(6.8)$$

which leads to the exact explicit formula

$$\Psi(x) = x - \log(2\pi) - \frac{1}{2}\log\left(1 - \frac{1}{x^2}\right) - \sum_{\zeta(\rho)=0} \frac{x^\rho}{\rho}. \qquad (6.9)$$

This is known as Mangoldt's formula (1895) and is one of the most important formulae in analytic theory of numbers. $\Psi(x)$ is real and gives the jumps for prime powers x. Although the last term looks complex, it is not, since the zeros enter pairwise and hence it is also real.

$\Psi(x)$ is equivalent to Riemann's $\Pi(x)$ and one has to admit that the formula for $\Psi(x)$ was deduced much more easily than the formula for $\Pi(x)$, with which we began this chapter. No wonder that it is meanwhile considered preferable to that of $\Pi(x)$.

Chapter 7
The Number of Roots in the Critical Strip

The following theorem was originally formulated by Riemann—but not proved. It was not until 1905 that von Mangoldt proved that the number of zeros of ζ in the critical range $0 < Re(s) < 1, 0 < t < T$ is given by

$$N(T) = \frac{T}{2\pi} \log \frac{T}{2\pi} - \frac{T}{2\pi}. \tag{7.1}$$

To prove this statement, let us assume $T \geq 3$ and $\zeta(s) \neq 0$ for $t = T$.
Then consider the rectangular R_T in the complex plane (Fig. 7.1):
    The zeros of the ξ function are identical to the ones of the ζ function in the critical range. Symmetry with respect to the axis $Re(s) = 1/2$ yields (remember from the logarithmic residue)

$$2N(T) = \frac{1}{2\pi i} \int_{\partial R_T} \frac{\xi'(s)}{\xi(s)} ds. \tag{7.2}$$

From the functional equation of ξ we obtain

$$\xi(1 - s) = \xi(s)$$
$$-\frac{\xi'(1 - s)}{\xi(1 - s)} = \frac{\xi'(s)}{\xi(s)}. \tag{7.3}$$

$C_T'(C_T)$ is the left (right) boundary of R_T:

$$\int_{C_T'} \frac{\xi'(s)}{\xi(s)} ds = \int_{C_T} \frac{\xi'(1 - s)}{\xi(1 - s)} d(1 - s) = \int_{C_T} \frac{\xi'(s)}{\xi(s)} ds$$

$$> N(T) = \frac{1}{2\pi i} \int_{C_T} \frac{\xi'(s)}{\xi(s)} ds. \tag{7.4}$$

© The Author(s), under exclusive license to Springer Nature Switzerland AG 2021
W. Dittrich, *Reassessing Riemann's Paper*,
SpringerBriefs in History of Science and Technology,
https://doi.org/10.1007/978-3-030-61049-4_7

Fig. 7.1 Boundary of R_T

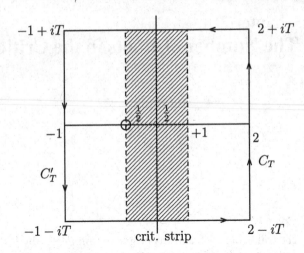

Now, using the following representation of the ξ function,

$$\xi(s) = \frac{s(s-1)}{2}\pi^{-s/2}\Gamma\left(\frac{s}{2}\right)\zeta(s) \tag{7.5}$$

we take the logarithm

$$\log\xi(s) = -\log 2 + \log s + \log(s-1) - \frac{s}{2}\log\pi + \log\Gamma\left(\frac{s}{2}\right) + \log\zeta$$

$$> \frac{d}{ds}\log\xi(s) = \frac{\xi'(s)}{\xi(s)} = \frac{1}{s} + \frac{1}{s-1} - \frac{1}{2}\log\pi + \frac{1}{2}\frac{\Gamma'\left(\frac{s}{2}\right)}{\Gamma\left(\frac{s}{2}\right)} + \frac{\zeta'(s)}{\zeta(s)}$$

$$> 2\pi i N(T) = \underbrace{\int_{C_T}\left(\frac{1}{s} + \frac{1}{s-1}\right)ds}_{①} - \underbrace{\int_{C_T}\frac{1}{2}\log\pi ds}_{②} + \underbrace{\frac{1}{2}\int_{C_T}\frac{\Gamma'\left(\frac{s}{2}\right)}{\Gamma\left(\frac{s}{2}\right)}ds + \int_{C_T}\frac{\zeta'(s)}{\zeta(s)}ds}_{③}$$

$$\tag{7.6}$$

① $\displaystyle\int_{C_T}\left(\frac{1}{s} + \frac{1}{s-1}\right)ds = \frac{1}{2}\int_{\partial R_T}\left(\frac{1}{s} + \frac{1}{s-1}\right)ds \overset{\text{resid}}{=} \frac{1}{2}2\pi i(1+1) = 2\pi i$

② $\displaystyle\int_{C_T}\frac{1}{2}\log\pi ds = \frac{1}{2}\log\pi\left(\left(\frac{1}{2}+iT\right) - \left(\frac{1}{2}-iT\right)\right) = iT\log\pi$

③ $\displaystyle\int_{C_T}\frac{1}{2}\frac{\Gamma'\left(\frac{s}{2}\right)}{\Gamma\left(\frac{s}{2}\right)}ds = \log\Gamma\left(\frac{s}{2}\right)\Big|_{\frac{1}{2}-iT}^{\frac{1}{2}+iT}$

$$= \log\Gamma\left(\frac{1}{4}+i\frac{T}{2}\right) - \log\Gamma\left(\frac{1}{4}-i\frac{T}{2}\right) \tag{7.7}$$

$$\log \Gamma(\bar{s}) = \overline{\log \Gamma(s)} := 2i \operatorname{Im} \log \Gamma \left(\frac{1}{4} + i \frac{T}{2} \right)$$

$$\underset{T \geq 3}{\overset{\text{Expand}}{=}} 2i \operatorname{Im} \left(\log \sqrt{2\pi} + \left(-\frac{1}{4} + i \frac{T}{2} \right) \log \left(i \frac{T}{2} \right) - i \frac{T}{2} + \mathcal{O} \left(\frac{1}{T} \right) \right)$$

$$= 2i \operatorname{Im} \left(\log \sqrt{2\pi} + \left(-\frac{1}{4} + i \frac{T}{2} \right) \left(\log \frac{T}{2} + i \frac{\pi}{2} \right) - i \frac{T}{2} + \mathcal{O} \left(\frac{1}{T} \right) \right)$$

$$= 2i\pi \left(\frac{T}{2\pi} \log \frac{T}{2} - \frac{T}{2\pi} \right) - \frac{1}{8} + \mathcal{O} \left(\frac{1}{T} \right). \tag{7.8}$$

Our intermediate result is then

$$2\pi i N(T) = 2\pi i - iT \log \pi + 2\pi i \left(\frac{T}{2\pi} \log \frac{T}{2} - \frac{T}{2\pi} - \frac{1}{8} + \mathcal{O} \left(\frac{1}{T} \right) \right) + \int_{C_T} \frac{\zeta'(s)}{\zeta(s)} ds. \tag{7.9}$$

$$\boxed{N(T) = 1 - \frac{T}{2\pi} \log \pi + \frac{T}{2\pi} \log \frac{T}{2} - \frac{T}{2\pi} - \frac{1}{8} + \mathcal{O} \left(\frac{1}{T} \right) + \frac{1}{2\pi i} \int_{C_T} \frac{\zeta'(s)}{\zeta(s)}.} \tag{7.10}$$

The last term can be split up into two parts, the results of which are given without further detailed calculations:

$$\int_{2-iT}^{2+iT} \frac{\zeta'(s)}{\zeta(s)} ds = \mathcal{O}(1), \quad \text{for } T \geq 3 \tag{7.11}$$

and using

$$\int_{\frac{1}{2}-iT}^{2-iT} \frac{\zeta'(s)}{\zeta(s)} ds = \int_{1/2}^{2} \frac{\zeta'(\sigma - iT)}{\zeta(\sigma - iT)} d\sigma = \int_{1/2}^{2} \overline{\frac{\zeta'(\sigma - iT)}{\zeta(\sigma + iT)}} ds$$

$$= \overline{\int_{\frac{1}{2}+iT}^{2+iT} \frac{\zeta'(s)}{\zeta(s)} ds}$$

$$> \frac{1}{2\pi i} \left(\int_{\frac{1}{2}-iT}^{2-iT} \frac{\zeta'(s)}{\zeta(s)} ds + \int_{2+iT}^{\frac{1}{2}+iT} \frac{\zeta'(s)}{\zeta(s)} ds \right) = \frac{1}{\pi} \operatorname{Im} \left(\int_{2+iT}^{\frac{1}{2}+iT} \frac{\zeta'(s)}{\zeta(s)} ds \right). \tag{7.12}$$

So far we have found

$$N(T) = \frac{T}{2\pi} \log \frac{T}{2\pi} - \frac{T}{2\pi} + \frac{7}{8} + \mathcal{O}\left(\frac{1}{T}\right) + \frac{1}{\pi} Im \left(\int\limits_{2+iT}^{\frac{1}{2}+iT} \frac{\zeta'(s)}{\zeta(s)} ds \right). \quad (7.13)$$

Using

$$\int\limits_{2+iT}^{\frac{1}{2}+iT} \frac{\zeta'(s)}{\zeta(s)} ds = \log \zeta \left(\frac{1}{2} + iT\right) - \log \zeta (2 + iT)$$

$$> Im \left(\int\limits_{2+iT}^{\frac{1}{2}+iT} \frac{\zeta'(s)}{\zeta(s)} ds \right) = arg \left(\zeta \left(\frac{1}{2} + iT\right) \right) - arg \left(\zeta (2 + iT) \right). \quad (7.14)$$

The modulus of the last expression can be shown to be $\mathcal{O}(\log T)$.

Hence our final result for the number of zeros in the critical strip with $0 < T$ is given by

$$\boxed{N(T) = \frac{T}{2\pi} \left(\log \frac{T}{2\pi} - 1 \right) + \mathcal{O}(\log T).} \quad (7.15)$$

As mentioned above, this formula was given by Riemann in 1859, but only proved by von Mangoldt in 1905.

By the way, we can also approximate $Im \log \Gamma(1/4 + it/2)$ and so obtain

$$Im \left\{ \log \Gamma \left(\frac{1}{4} + \frac{it}{2} \right) \right\} = \frac{t}{2} \log \left(\frac{t}{2}\right) - \frac{t}{2} - \frac{\pi}{8} - \frac{t}{2} \log \pi + \mathcal{O}(t^{-1})$$

$$\text{i.e.} \quad \vartheta(t) = \frac{t}{2} \log \left(\frac{t}{2\pi}\right) - \frac{t}{2} - \frac{\pi}{8} + \mathcal{O}(t^{-1}). \quad (7.16)$$

This brings us to the useful result

$$N(T) = \frac{1}{\pi} \vartheta(T) + 1 + \frac{1}{\pi} arg \zeta \left(\frac{1}{2} + iT\right), \quad (7.17)$$

with

$$\frac{1}{\pi} arg \zeta \left(\frac{1}{2} + iT\right) = \mathcal{O}(\log T) \quad \text{for} \quad T \to \infty. \quad (7.18)$$

So we can conclude for the number of zeros of ζ in the critical strip:

$$1. N(T) \overset{T \to \infty}{\longrightarrow} \infty$$

$$2. N(T) \sim \frac{T}{2\pi} \log T. \tag{7.19}$$

This follows from

$$N(T) = \frac{T}{2\pi} \log \frac{T}{2\pi} + \mathcal{O}(\log T), \tag{7.20}$$

which when divided by $T/2\pi \log T$, leads to

$$\frac{N(T)}{\frac{T}{2\pi} \log T} = \frac{\log T - \log 2\pi}{\log T} + \frac{C}{T/2\pi} \underset{T \to \infty}{\longrightarrow} 1. \tag{7.21}$$

This result should be compared with the prime number theorem (Gauß 1796, when he was 15 years old)

$$\pi(x) \sim \frac{x}{\log x} \quad \text{or} \quad \lim_{x \to \infty} \left(\frac{\pi(x)}{\frac{x}{\log x}} \right) = 1. \tag{7.22}$$

Von Koch proved in 1901: If the Riemann hypothesis $\left(Re(s) = \frac{1}{2} \right)$ is true, then

$$\pi(x) = Li(x) + \mathcal{O}\left(\sqrt{x} \log x \right), \tag{7.23}$$

i.e., the error in the claim $\pi(x) \sim Li(x)$ is of the order $\sqrt{x} \log x$.

Chapter 8
Riemann's Zeta Function Regularization

In this section, we want to introduce the concept of the zeta function in connection with regularizing certain problems in quantum physics where infinities occur. For this reason, we consider an operator A with positive, real discrete eigenvalues $\{a_n\}$, i.e., $A f_n(x) = a_n f(x)$ and one defines its associated zeta function by

$$\zeta_A(s) = \sum_n a_n^{-s} = \sum_n e^{-s \ln a_n}, \tag{8.1}$$

where n runs over all eigenvalues. If one chooses for A the Hamilton operator of the harmonic oscillator, for example, one gets (apart from the zero-point energy) exactly the Riemann zeta function. By formal differentiation now follows:

$$\zeta_A'(0) = - \sum_n \ln a_n e^{-s \ln a_n}\bigg|_{s=0} = - \ln \left(\prod_n a_n \right). \tag{8.2}$$

This suggests the definition

$$\det A = \exp\left[-\zeta_A'(0)\right], \tag{8.3}$$

which we shall exclusively be using in the following. The advantage of this method is that $\zeta_A'(0)$ is not singular for many operators of physical interest. As an example of the many applications to relativistic as well as non-relativistic problems in quantum field theory, we will choose the Casimir effect.

This effect is a non-classical electromagnetic, attractive or repulsive force which occurs between electrically neutral conductors in a vacuum. The size of this force was first calculated by Casimir for the case of ideal conducting, infinitely extended, parallel plates; his result was a force

© The Author(s), under exclusive license to Springer Nature Switzerland AG 2021
W. Dittrich, *Reassessing Riemann's Paper*,
SpringerBriefs in History of Science and Technology,
https://doi.org/10.1007/978-3-030-61049-4_8

$$F = -\frac{\pi^2}{240} \cdot \frac{\hbar c}{a^4}, \tag{8.4}$$

where a is the distance between the plates and the negative sign indicates that the plates attract each other. This force apparently depends only on the fundamental constants $\hbar$ and c apart from the distance between the plates; not, however, on the coupling constant α between the Maxwell and the matter field. Its quantum mechanical character is revealed by the fact that F vanishes in the classical limit $\hbar \rightarrow 0$.

Casimir's derivation of F was based on the concept of a quantum electrodynamic (particle) vacuum representing the zero-point oscillations of an infinite number of harmonic oscillators. As a result, one gets the total vacuum energy by summation over the zero-point energies $1/2\hbar\omega_{\vec{k}}$ of all allowed modes with wave number vector $\vec{k}$ and polarization σ,

$$E = \sum_{\vec{k},\sigma} \frac{1}{2}\hbar\omega_{\vec{k}}. \tag{8.5}$$

If we evaluate this equation for the case of two plane parallel plates at distance a from each other, one does get a divergent total energy $E(a)$, but the energy difference $E(a) - E(a + \delta a)$ is finite (δa = infinitesimal change in the plate distance), leading also to a finite force per unit area,

$$F = -\frac{\partial E(a)}{\partial a}. \tag{8.6}$$

To calculate this energy difference or force, a UV-cut-off is usually introduced, i.e., the energy E is replaced by

$$\sum_{\vec{k},\sigma} \frac{1}{2}\hbar\omega_{\vec{k}}e^{-\frac{b}{\pi c}\omega_{\vec{k}}} \tag{8.7}$$

and, in the end result, the limit $b \rightarrow 0$ is considered.

This derivation of F, however, can give the impression that the appearance of the Casimir force is linked to the existence of the zero-point fluctuations of the quantized electromagnetic field.

In order to avoid the divergent vacuum energy problem, in the following, we shall consider the problem according to Hawking from the viewpoint of path integral quantization and zeta-function regularization. Here, it is again unnecessary to refer to the vacuum oscillation. For reasons of simplicity, we wish to consider the Casimir effect only for a real, scalar field theory which is defined by ($\hbar = c = 1!$)

$$\mathcal{L}(\phi) = -\frac{1}{2}\partial_\mu\phi\partial^\mu\phi - \frac{1}{2}m^2\phi^2 - V(\phi), \tag{8.8}$$

with the arbitrary potential V.

First, we couple the field ϕ to an external source J,

$$\mathcal{L}(\phi) \to \mathcal{L}(\phi) + J\phi. \tag{8.9}$$

We can then write the vacuum amplitude $\langle 0_+ | 0_- \rangle^J$ or the action $W[J]$ in the form

$$\langle 0_+ | 0_- \rangle^J = e^{iW[J]} = \int [d\phi] e^{i \int d^4 x \{\mathcal{L}(\phi) + J\phi\}}, \tag{8.10}$$

where we guarantee the convergence of the path integral by the substitution $m^2 \to m^2 - i\epsilon,\ \epsilon > 0$. We have assumed that $|0_-\rangle$ or $|0_+\rangle$ describes a vacuum which is not "disturbed" by the presence of certain geometries, i.e., the path integral is, without restriction by boundary conditions, to be taken over all fields ϕ. This changes as soon as we introduce two plates into the vacuum, for example, perpendicular to the z axis (points of intersection: $z = 0$ and $z = a$) and require that only those fields should contribute to the path integral which would vanish on the plate surface, i.e., for which it holds that

$$\phi(x_0, x_1, x_2, 0) = \phi(x_0, x_1, x_2, a) = 0 \tag{8.11}$$

for arbitrary (x_0, x_1, x_2). We now get

$$\langle 0_+ | 0_- \rangle_a^J = e^{iW(a,[J])}$$

$$= \int_{\mathcal{F}_a} [d\phi] \exp\left[i \int d^4 x \left\{ -\frac{1}{2} \partial_\mu \phi \partial^\mu \phi - \frac{1}{2}(m^2 - i\epsilon)\phi^2 - V(\phi) - J\phi \right\} \right], \tag{8.12}$$

where $\int_{\mathcal{F}_a}$ suggests that the path integral is only to be taken over the restricted space of functions $\mathcal{F}_a$ defined by the boundary conditions. With this, we have represented the vacuum amplitude or the action for the most general case as a function of the geometric parameter a and as a functional of the external source J. In order to approach the conditions of the QED Casimir effect, we now choose $J = 0$ as well as a free ($V = 0$), massless ($m = 0$) field ϕ. Following a partial integration:

$$\langle 0_+ | 0_- \rangle_a = e^{iW(a)} = \int_{\mathcal{F}_a} [d\phi] e^{-\frac{i}{2} \int d^4 x \phi \{-\partial^2 - i\epsilon\}\phi}. \tag{8.13}$$

The Gauss integral gives

$$\langle 0_+ | 0_- \rangle_a = e^{iW(a)} = \int_{\mathcal{F}_a} [d\phi] e^{-\frac{1}{2} \int d^3 x d\tau \phi \{-\Box_E\}\phi}. \tag{8.14}$$

Here, N is a (divergent) constant which we shall set $= 1$, since it only contributes a non-physical additive constant to $W(a)$. By writing $\Box_E/\mathcal{F}_a$, we mean that only eigenvalues with eigenfunctions in $\mathcal{F}_a$ can be used to evaluate the determinant. Furthermore (in keeping with the $i\epsilon$ requirement), a Wick rotation $t \rightarrow i\tau$ was made, i.e., $\Box_E = \partial_\tau^2 + \Delta$.

From the original definition of the determinant, it follows that

$$\langle 0_+|0_-\rangle_a = e^{iW(a)} = \left[\exp\left\{-\zeta'_{-\Box_E/\mathcal{F}_a}(0)\right\}\right]^{-\frac{1}{2}}$$

$$= \exp\left[\frac{1}{2}\zeta'_{-\Box_E/\mathcal{F}_a}(0)\right]. \tag{8.15}$$

The operator $-\Box_E/\mathcal{F}_a$ has the spectrum

$$\left\{k_0^2 + k_1^2 + k_2^2 + \left(\frac{\pi n}{a}\right)^2 |k_0, k_1, k_2 \in \mathbb{R}, n \in \mathbb{N}\right\} \tag{8.16}$$

and thus, the zeta function

$$\zeta_{-\Box_E/\mathcal{F}_a}(s) = 2\frac{A}{(2\pi)^2}\frac{T_E}{2\pi}\int\limits_{-\infty}^{\infty}\int\int dk_0 dk_1 dk_2 \sum_{n=1}^{\infty}\left[k_0^2 + k_1^2 + k_2^2 + \left(\frac{n\pi}{a}\right)^2\right]^{-s}. \tag{8.17}$$

Here, the factor 2 makes allowance for the two polarization possibilities of the photon, which, in our simple model, have no analogue. Furthermore, AT_E is a normalization volume in three-dimensional $(0, 1, 2)$ space, where the Euclidean time T_E is linked to a (Minkowski) normalization time interval T by $T_E = iT$. Dropping the term independent of a ($n = 0$) in the last equation simply leads to the subtraction of an (infinite) constant of $W(a)$.

Further evaluation of $\zeta_{-\Box_E/\mathcal{F}_a}(s)$ now takes on the form

$$\zeta_{-\Box_E/\mathcal{F}_a}(s) = 2AT_E\frac{4\pi}{(2\pi)^3}\sum_{n=1}^{\infty}\int\limits_0^{\infty} dk k^2 \left[k^2 + \left(\frac{n\pi}{a}\right)^2\right]^{-s}$$

$$= \frac{8\pi}{(2\pi)^3}AT_E\left(\frac{\pi}{a}\right)^{3-2s}\sum_{n=1}^{\infty}n^{3-2s}\frac{1}{2}\frac{\Gamma\left(\frac{3}{2}\right)\Gamma\left(s - \frac{3}{2}\right)}{\Gamma(s)}$$

$$= \frac{4\pi}{(2\pi)^3}AT_E\left(\frac{\pi}{a}\right)^{3-2s}\zeta(2s-3)\frac{\Gamma\left(\frac{3}{2}\right)\Gamma\left(s - \frac{3}{2}\right)}{\Gamma(s)}. \tag{8.18}$$

The derivative is

$$\zeta'_{-\Box_E/\mathcal{F}_a}(0) = \frac{4\pi}{(2\pi)^3} A T_E \left(\frac{\pi}{a}\right)^3 \zeta(-3)\Gamma\left(\frac{3}{2}\right)\Gamma\left(-\frac{3}{2}\right)\frac{d}{ds}\frac{1}{\Gamma(s)}\bigg|_{s=0}$$

$$= \frac{\pi^2}{360a^3} A T_E. \tag{8.19}$$

Finally we get

$$\langle 0_+|0_-\rangle = e^{iW(a)} = e^{-\epsilon(a)T_E} = e^{-i\epsilon(a)T}, \tag{8.20}$$

with

$$\epsilon(a) = -\frac{\pi^2}{720a^3} A. \tag{8.21}$$

The appearance of the phase factor $e^{-i\epsilon(a)T}$ in the vacuum amplitude allows us to identify $\epsilon(a)$ as the vacuum energy displacement and to write, for the force per surface unit,

$$F = -\frac{1}{A}\frac{\partial\epsilon}{\partial a}, \tag{8.22}$$

which leads to

$$F = -\frac{\pi^2}{240}\cdot\frac{1}{a^4} \tag{8.23}$$

or, after putting $\hbar$ and c back in:

$$F = -\frac{\pi^2}{240}\cdot\frac{\hbar c}{a^4}. \tag{8.24}$$

This is precisely Casimir's result which we have now completely derived with the aid of Riemann's zeta-function regularization, which completely eliminated the divergent zero-point energy. The same procedure finds application in QED and QCD, and can be looked up in the list of references (i.e., in [10–12]).

Chapter 9
ζ-Function Regularization of the Partition Function of the Harmonic Oscillator

We start by making the following changes from Minkowski real time $t = x_0$ to Euclidean imaginary "time" $\tau = t_E$:

$$\tau = it = \beta. \tag{9.1}$$

Here we put $\hbar = 1$, T = temperature of the system = β^{-1}, k = 1. Then we write for the partition function of our one-dimensional quantum mechanical system:

$$Z = \int_{\substack{x(0) = x(\beta) \\ arbitr.}} [dx(\tau)] exp\left\{-\int_0^\beta d\tau\left[\frac{m}{2}\dot{x}^2(\tau) + V(x(\tau))\right].\right\} \tag{9.2}$$

The exponential is obtained from

$$iS = i\int_0^t dt'[\frac{m}{2}\dot{x}^2(t') - V(x(t'))], t' = -i\tau, \frac{d}{dt'} = -\frac{d}{id\tau}$$

$$0 \le t' \le t,$$

$$= -\int_0^\beta d\tau[\frac{m}{2}\dot{x}^2(\tau) + V(x(\tau))].$$

$$0 \le \tau \le \beta$$

W. Dittrich, *Reassessing Riemann's Paper*,
SpringerBriefs in History of Science and Technology,
https://doi.org/10.1007/978-3-030-61049-4_9

In the functional integration we require $x(\tau)$ to be periodic with period $\beta : x(\tau) = (\tau + \beta)$ or, if we put $\tau = 0$,

$$x(0) = x(\beta).$$

In particular, we obtain for the harmonic oscillator ($m = 1$), where

$$V(x) = \frac{\omega^2}{2} x^2,$$

$$Z = \int_{\beta} [dx(\tau)] exp \left\{ -\frac{1}{2} \int_0^{\beta} d\tau [\dot{x}^2(\tau) + \omega^2 x^2(\tau)] \right\},$$

or, performing an integration by parts on the first term in the exponential,

$$\frac{dx}{d\tau}\frac{dx}{d\tau} \rightarrow \overbrace{\frac{x(\tau)\dot{x}(\tau)]_0^{\beta}}{\text{per. func.} \rightarrow 0}} - x(\tau)\frac{d^2}{d\tau^2}x(\tau)$$

$$Z = \int_{\beta} [dx(\tau)] exp \left\{ -\frac{1}{2} \int_0^{\beta} d\tau x(\tau) \left[-\frac{d^2}{d\tau^2} + \omega^2 \right]_{\beta} x(\tau). \right. \tag{9.3}$$

The subscript β in $[-\frac{d^2}{d\tau^2} + \omega^2]_{\beta}$ indicates that the differential operator is restricted to the function space defined by $x(0) = x(\beta)$;

$$\Omega_{\tau} := [-\frac{d^2}{d\tau^2} + \omega^2]_{\beta}. \tag{9.4}$$

Ω_{τ} is a positive definite elliptic operator acting on $x_n(\tau)$ defined on a compact circular τ-manifold. Ω_{τ} has a complete set of orthonormal (real) eigenfunctions $x_n(\tau)$ and associated eigenvalues λ_n:

$$\left[-\frac{d^2}{d\tau^2} + \omega^2 \right]_{\beta} x_n(\tau) = \lambda_n x_n(\tau), \tag{9.5}$$

with

$$x_n(0) = x_n(\beta), \quad \int_0^{\beta} d\tau x_m(\tau) x_n(\tau) = \delta_{mn}. \tag{9.6}$$

Explicitly,

$$x_n(\tau) = \begin{cases} \sqrt{\frac{2}{\beta}} \sin \frac{2\pi}{\beta} n\tau, \\ \sqrt{\frac{2}{\beta}} \cos \frac{2\pi}{\beta} n\tau, \end{cases}$$

$$\boxed{\lambda_n = \left(\frac{2\pi n}{\beta}\right)^2 + \omega^2, n \in Z} \qquad (9.7)$$

Check: $\frac{d^2}{d\tau^2}\sqrt{\frac{2}{\beta}}\sin\frac{2\pi}{\beta}n\tau = -(\frac{2\pi}{\beta}n)^2\sqrt{\frac{2}{\beta}}\sin\frac{2\pi}{\beta}n\tau,$

$$\left[-\frac{d^2}{d\tau^2} + \omega^2\right]\sqrt{\frac{2}{\beta}}\sin\frac{2\pi}{\beta}n\tau = \left[\left(\frac{2\pi}{\beta}n\right)^2 + \omega^2\right]\sqrt{\frac{2}{\beta}}\sin\frac{2\pi}{\beta}n\tau = \lambda_n\sqrt{\frac{2}{\beta}}\sin\frac{2\pi}{\beta}n\tau.$$

Periodicity is also obvious: $x_n(\tau + \beta) = x_n(\tau)$,

and orthonormality follows from

$$\frac{2}{\beta}\int_0^\beta d\tau \cos\frac{2\pi}{\beta}m\tau \cos\frac{2\pi}{\beta}n\tau = \delta mn.$$

So we can write

$$Z = \int_\beta [dx(\tau)]e^{-S_E[x(\tau)]} \qquad (9.8)$$

with the Euclidean action

$$S_E[x(\tau)] = \frac{1}{2}\int_0^\beta d\tau x(\tau)\Omega_\tau x(\tau), \qquad (9.9)$$

or

$$Z = (\det \Omega_\tau)^{-\frac{1}{2}}, \qquad \Omega_\tau = \left[-\frac{d^2}{d\tau^2} + \omega^2\right]_\beta. \qquad (9.10)$$

Here again, we use the ζ-function evaluation of an operator A:

$$\det A = e^{-\zeta_A'(0)} \qquad (9.11)$$

or

$$ln \det A = -\zeta_A'(0).\tag{9.12}$$

Now let us return to (9.10) and apply (9.12):

$$\ln Z = -\frac{1}{2}ln \det \Omega = \frac{1}{2}\zeta_\Omega'(0),\tag{9.13}$$

where $\zeta_\Omega(s) = \sum_{n\in Z} \lambda_n^{-s}$.

with

$$\lambda_n = (\frac{2\pi n}{\beta})^2 + \omega^2, \qquad n \in Z.\tag{9.14}$$

Expression (9.14) is also known as the "Matsubara rule" for Bose particles. In order to evaluate (9.13), we have to discuss the thermal ζ-function,

$$\zeta_\Omega(s) = \sum_{n=-\infty}^{+\infty}\left[\left(\frac{2\pi}{\beta}\right)^2 n^2 + \omega^2\right]^{-s} = \left(\frac{2\pi}{\beta}\right)^{-2s}\sum_{n=-\infty}^{+\infty}\left[n^2 + \left(\frac{\omega\beta}{2\pi}\right)^2\right]^{-s}$$

$$=: \left(\frac{\beta}{2\pi}\right)^{2s} D\left(s, \frac{\omega\beta}{2\pi}\right)$$

or

$$\zeta_\Omega(s) = \left(\frac{\beta}{2\pi}\right)^{2s} D(s, v),\tag{9.15}$$

where

$$D(s, v) := \sum_{n=-\infty}^{+\infty}(n^2 + v^2)^{-s} \qquad v := \frac{\omega\beta}{2\pi}.\tag{9.16}$$

The series (9.16) converges for $Re s > \frac{1}{2}$, and its analytic continuation defines a meromorphic function of s, analytic at $s = 0$. Further properties are:

$$D(0, v) = 0,\tag{9.17}$$

which implies $\zeta_\Omega(0) = 0$, and since

$$\det(\mu\Omega) = \mu^{\zeta_\Omega(0)}\det\Omega = \det\Omega,\tag{9.18}$$

there is no scale change.

The first equal sign in (9.18) follows from

$$\det(\mu\Omega) = e^{-\zeta'_{\mu\Omega}(0)} = e^{-\frac{d}{ds}\left(\zeta_{\mu\Omega}(s)\right)}|s = 0$$

Here we need

$$\zeta_{\mu\Omega}(s) = \sum_n (\mu\lambda_n)^{-s} = \mu^{-s}\sum_n \lambda_n^{-s} = \mu^{-s}\zeta_\Omega(s)$$
$$= e^{-s\,\ln\mu}\zeta_\Omega(s),$$

which yields

$$\frac{d}{ds}\zeta_{\mu\Omega}(s) = -\ln\mu e^{-s\,\ln\mu}\zeta_\Omega(s) + e^{-s\ln\mu}\zeta'_\Omega(s)$$

and, putting $s = 0$,

$$-\frac{d}{ds}\zeta_{\mu\Omega}(s)|_{s=0} = \ln\mu\zeta_\Omega(0) - \zeta'_\Omega(0),$$

we obtain indeed

$$e^{-\zeta'_{\mu\Omega}(0)} = \mu^{\zeta_\Omega(0)}e^{-\zeta'_\Omega(0)} = e^{-\zeta'_\Omega(0)}.$$

In addition to (9.16), we list the following properties:

$$\frac{\partial}{\partial s}D(s,v)|_{s=0} = -2\ln(2\sinh\pi v),$$

$$D(1,v) = \frac{\pi}{v}\cos h\pi v,$$

$$D\left(-\frac{1}{2},0\right) = -\frac{1}{6}, \quad D\left(-\frac{3}{2},0\right) = \frac{1}{60}.$$

$D(s,v)$ has poles at $s = \frac{1}{2}, -\frac{1}{2}, -\frac{3}{2}, -\frac{5}{2},...$
We are interested in

$$\zeta'_\Omega(s)|_{s=0} = \frac{d}{ds}(\frac{\beta}{2\pi})^{2s}|_{s=0}\overbrace{D(0,v)}^{=0} + (\frac{\beta}{2\pi})^{2s}|_{s=0}\frac{d}{ds}D(s,v)|_{s=0}$$
$$= -2\ln(2\sin h\pi v) = -2ln(2\sin h\frac{\omega\beta}{2})$$
$$= -2\left[\ln 2 + \ln\left(\sin h\frac{\omega\beta}{2}\right)\right].$$

Setting $x; = \frac{\omega\beta}{2}$, we can continue to write

$$\zeta'_\Omega(s)|_{s=0} = -2\left[\ln 2 + \ln\left(\frac{e^x - e^{-x}}{2}\right)\right]$$
$$= -2\left[+\ln(e^x(1 - e^{-2x}))\right]$$
$$= -2\left[x + \ln(1 - e^{-2x})\right].$$

So we finally obtain

$$\zeta'_\Omega(s)|_{s=0} = -\omega\beta - 2\ln(1 - e^{-\omega\beta})$$

or, according to (9.13),

$$\ln Z = -\frac{1}{2}\zeta'_\Omega(0) = -\frac{1}{2}\omega\beta - \ln(1 - e^{-\omega\beta}). \tag{9.19}$$

Note that we have clearly isolated the zero-point energy.
Another useful form is

$$\ln Z = \frac{1}{2}\zeta'_\Omega(0) = -\ln\left(2\sin h\frac{\omega\beta}{2}\right),$$

which implies

$$\boxed{Z = \frac{1}{2\sin h\left(\frac{\omega\beta}{2}\right)}} \tag{9.20}$$

Chapter 10
ζ-Function Regularization of the Partition Function of the Fermi Oscillator

Here we start with the action for the Fermi oscillator:

$$S[x(t)] = \frac{i}{2} \int_0^t dt' [x^T \dot{x} + \omega x^T M x] \tag{10.1}$$

$$x = \begin{pmatrix} x_1 \\ x_2 \end{pmatrix}, M = \begin{pmatrix} 0 & 1 \\ -1 & 0 \end{pmatrix}, = \frac{i}{2} \int_0^t dt' x^T \left(\frac{\partial}{\partial t} + i\omega\sigma_2 \right) x \tag{10.2}$$

$$iM = -\sigma_2 \ A := \partial/\partial t + i\omega\sigma_2 \ (x \equiv)\psi = \frac{1}{\sqrt{2}}(\psi_1 - i\psi_2)$$

$$\psi^T = \psi^* = (\psi_1 + i\psi_2)$$

$$x_i = a\,numbers! = \psi_i$$
$$\psi_i^2 = 0$$

The operator $A = \partial/\partial t + i\omega\sigma_2$ is anti-Hermitian with respect to the scalar product

$$(x, y) = \int_{-\infty}^{+\infty} dt\, x^\dagger(t) y(t).$$

($\dagger = T$, since x_i, y_i are real)

$$(x, A\,y) = \int dt\, x^\dagger(t) A\, y(t)$$

© The Author(s), under exclusive license to Springer Nature Switzerland AG 2021
W. Dittrich, *Reassessing Riemann's Paper*,
SpringerBriefs in History of Science and Technology,
https://doi.org/10.1007/978-3-030-61049-4_10

$$= \int dt\, x^\dagger(t) \left[\frac{\vec{\partial}}{\partial t} + i\omega\sigma_2\right] y(t) = \int dt\, x^\dagger(t) \left[-\frac{\overleftarrow{\partial}}{\partial t} + i\omega\sigma_2\right] y(t) .$$

int b.p.

$$\sigma_2^\dagger = \sigma_2 = \int dt \left[-\left(\frac{\partial}{\partial t} + i\omega\sigma_2\right)x(t)\right]^\dagger y(t) = -(Ax, y)$$

$$> e^{iS} = e^{i\frac{1}{2}\int\limits_0^t dt' x^T Ax} \tag{10.3}$$

$$> \int \mathcal{D}x_1(t)\mathcal{D}x_2(t) e^{-\frac{1}{2}\int\limits_0^t dt' x^T Ax} = det_{\sigma,t} A \tag{10.4}$$

Consider $det\left[\frac{\partial}{\partial t} + i\omega\sigma_2\right] = det[\sigma_1^2(\frac{\partial}{\partial t} + i\omega\sigma_2)] = \qquad \sigma_1^2 = 1$

$$\{\sigma_1, \sigma_2\} = 0 \qquad = det\left[\sigma_1\left(\frac{\partial}{\partial t} + i\omega\sigma_2\right)\sigma_1\right] = det[(\frac{\partial}{\partial t} - (i\omega\sigma_2)\sigma_1^2]$$

$$= det[\frac{\partial}{\partial t} - i\omega\sigma_2]$$

$$> det\left(\frac{\partial}{\partial t} + i\omega\sigma_2\right) = det\left(\frac{\partial}{\partial t} - i\omega\sigma_2\right).$$

Now use $det\, M = e^{tr\ln M}$ to write

$$ln\ det(\partial_t + i\omega\sigma_2) = trln(\partial_t + i\omega\sigma_2)$$
$$= tr\, ln[(\partial_t + i\omega\sigma_2)(\partial_t - i\omega\sigma_2)]$$
$$= tr\, ln(\partial_t^2 + \omega^2) - tr\, ln(\partial_t - i\omega\sigma_2)$$
$$tr\, 1_2 = 2 := 2tr_t ln(\partial_t^2 + \omega^2) - tr\, ln(\partial_t - i\omega\sigma_2)$$
$$> ln\, det(\partial_t + i\omega\sigma_2) + ln\, det(\partial_t - i\omega\sigma_2) = 2tr_t\, ln(\partial_t^2 + \omega^2)$$
$$> 2ln\, det(\partial_t + i\omega\sigma_2) = 2tr_t\, ln(\partial_t^2 + \omega^2)$$
$$> e^{tr\, ln(\partial_t + i\omega\sigma_2)} = e^{tr_t\, ln(\partial_t^2 + \omega^2)}$$
$$> det_{\sigma,t}\left(\frac{\partial}{\partial t} + i\omega\sigma_2\right) = det_t(\partial_t^2 + \omega^2).$$

Finally, $\boxed{\int \mathcal{D}x(t) e^{-\frac{1}{2}\int\limits_0^t dt' x^T Ax} = det_t\left(\frac{\partial^2}{\partial t^2} + \omega^2\right).}$

Matsubara's rule yields

$$\lambda_n = \left(\frac{(2\pi)}{\beta}\right)^2 \left(n + \frac{1}{2}\right)^2 + \omega^2 \tag{10.5}$$

$$Z^{Fermi} = \prod_{n=0}^{\infty} \left(\frac{(2\pi)}{\beta}\right)^2 \left(n + \frac{1}{2}\right)^2 + \omega^2 = det\,A \tag{10.6}$$

instead of

$$Z^{Bose} = \prod_{n=-\infty}^{\infty} \left(\frac{(2\pi)}{\beta}\right) n^2 + \omega^2 = (det\,\Omega_\tau)^{-\frac{1}{2}} \tag{10.7}$$

To obtain the ζ-function, replace λ_n in $Z^{Fermi} = \prod_{n=0}^{\infty} \lambda^n$ by λ_n^{-s} and take the sum instead of the product:

$$\zeta^F(s) = \sum_{n=0}^{\infty} [(2n+1)^2 \frac{\pi^2}{\beta^2} + \omega^2]^{-s} \qquad \begin{array}{l} 2n+1 = m \\ n = 0, 1, 2 \\ m = 1, 3, 5 \end{array}$$

$$= \sum_{m=1,3\ldots} [m^2 \frac{\pi^2}{\beta^2} + \omega^2]^{-s}$$

$$= \left(\frac{\pi}{\pi\beta}\right)^{-2s} \sum_{m=1,3\ldots} [m^2 + \left(\frac{\omega\beta}{\pi}\right)^2]^{-s}$$

$$= \left(\frac{\beta}{\pi}\right)^{2s} \frac{1}{2} \sum_{m=\pm 1, \pm s, \ldots} [m^2 + \left(\frac{\omega\beta}{\pi}\right)^2]^{-s}$$

$$= \frac{1}{2} \left(\frac{\beta}{\pi}\right)^{2s} \sum_{n=-\infty}^{+\infty} \left\{ [n^2 + \left(\frac{\omega\beta}{\pi}\right)^2]^{-s} - [4n^2 + \left(\frac{\omega\beta}{\pi}\right)^2]^{-s} \right\}$$

$$\begin{array}{ll} n = 0 & 0 \qquad \frac{\omega\beta}{\pi} = 2v \\ n = \pm 1 & (1 + (2v)^2)^{-s} - (4 + (2v)^2)^{-s} \\ n = \pm 2 & (4 + (2v)^2)^{-s} - (16 - (2v)^2)^{-s} \\ n = \pm 3 & (9 + (2v)^2)^{-s} - (36 - (2v)^2)^{-s} \\ n = \pm 4 & (16 + (2v)^2)^{-s} - \ldots \end{array}$$

$$\zeta^F(s) = \frac{1}{2} \left(\frac{\beta}{\pi}\right)^{2s} \left[(D(s, 2v) - 4^{-s} D(s, v)\right] \quad \zeta_F(0) = 0$$
$$= 0 \qquad = 0$$

$$\zeta^{F'}(s)|_{s=0} = \frac{1}{2} \frac{d}{ds} \left(\left(\frac{\beta}{\pi}\right)^{2s}|_{s=0}\right) [D(0.2v) - D(0, v)] +$$

$$+ \frac{1}{2} [\frac{d}{ds} D(s, 2v)|_0 - \frac{d}{ds} (4^{-s} D(s, v)|_0]$$

$$- 2ln(2 \sin h\pi v) \quad \{-s4^{-s-1} D(s, v)|_0\}$$

$$+ 4^{-s}\left[\frac{d}{ds}D(s, v)\right]|_0\Bigg\}$$

$$= -2ln(2\sin h\pi v)$$

$$= -ln(2\sin h2\pi v) + ln(2\sin h\pi v)$$

$\zeta_\Omega(0)\, index\ of\ \Omega$

$\zeta_\Omega^{B,F}(0) = 0$

$\Omega_F =:= A\ \ top.invar.$

$\det(\mu\Omega) = \mu^{-\zeta_\Omega(0)}det\Omega$

$lnZ^F = \ln\det\Omega_F$

$>\ Z^{B,F}$ are independent of scale appearing in the problem

$= \ln(e^{-\zeta_F'(0)})$

$= -\zeta_F'(0)$

$= ln\left(\dfrac{\sin h2\pi v}{\sin h\pi v}\right) = ln[2(\cos h\pi v)]$

$Z^F = 2\cos h\pi v$

$$\boxed{Z^F = 2\cos h\frac{\beta\omega}{2}}$$

From the former results it is easy to obtain the supersymmetric Bose-Fermi oscillator at fine temperature T.

Partition functions:

$$Z^B = \frac{1}{2\sin h\left(\frac{\beta\omega}{2}\right)},\ Z^F = 2\cos h\left(\frac{\beta\omega}{2}\right)$$

$$Z \equiv Z^{SUSY} = Z^B Z^F = \cot h\left(\frac{\beta\omega}{2}\right)$$

Compute:

$$\frac{\partial Z^B}{\partial\beta} = \frac{1}{2}\frac{\partial}{\partial\beta}\frac{1}{\sin h\left(\frac{\beta\omega}{2}\right)} = \frac{1}{2}(-)\frac{\cos h\left(\frac{\beta\omega}{2}\right)\frac{\omega}{2}}{sin^2 h\left(\frac{\beta\omega}{2}\right)}$$

$$= -\frac{\omega}{4}\frac{1}{\sin h\frac{\beta\omega}{2}}\cot h(\frac{\beta\omega}{2})$$

$$= -\frac{\omega}{2} Z^B \cot h\left(\frac{\beta\omega}{2}\right)$$

$$\frac{\partial (Z^B)^{-1}}{\partial\beta} = 2\frac{\omega}{2}\cos h(\frac{\beta\omega}{2}) = \frac{\omega}{2}Z^F$$

$$\frac{\partial Z^F}{\partial\beta} = 2\frac{\omega}{2}\sin h(\frac{\beta\omega}{2}) = \frac{\omega}{2}(Z^B)^{-1}$$

Facts:

$$Z^B = Tr\left(e^{-\beta H^B}\right)$$

$$\langle A\rangle = Tr(\varrho A), \quad \varrho = \frac{1}{Z}e^{-\beta E}$$

$$= \frac{1}{Z}Tr\langle Ae^{-\beta E}\rangle$$

$$-\frac{\omega}{2}Z^B \cot h(\frac{\beta\omega}{2}) = \frac{\partial Z^B}{\partial\beta} = -Tr\left(H^B e^{-\beta H^B}\right) = -Z^B\langle H^B\rangle$$

$$\langle H^B\rangle = \frac{\omega}{2}\cot h\left(\frac{\beta\omega}{2}\right)$$

$$\frac{\partial Z^F}{\partial\beta} = -Z^F\langle H^F\rangle = \frac{\omega}{2}(Z^B)^{-1}$$

$$\langle H^F\rangle = -\frac{\omega}{2}\frac{1}{Z^F Z^B} = -\frac{\omega}{2}\frac{1}{\cot h\left(\frac{\beta\omega}{2}\right)}$$

$$= -\frac{\omega}{2}\tan h\left(\frac{\beta\omega}{2}\right)$$

$$> Z^{SUSY} = \cot h\left(\frac{\beta\omega}{2}\right) = \frac{2}{\omega}\langle H^B\rangle$$

$$= \left(-\frac{2}{\omega}\langle H^F\rangle\right)^{-1}$$

$$\langle H^B\rangle\langle H^F\rangle = -\frac{\omega^2}{4}, \forall T$$

$$\beta = \frac{1}{kT}, T = 0: Z^{SUSY} = 1, \langle H^B\rangle_{T=0} = \frac{\omega}{2} = -\langle H^F\rangle_{T=0}$$

$$T \gg \omega : \langle H^B\rangle = -\frac{1}{\beta}\left(\frac{\beta\omega}{2}\right)\cot h\left(\frac{\beta\omega}{2}\right) = \frac{1}{\beta}(1 + \frac{1}{3}\left(\frac{\beta\omega}{2}\right)^2 \ldots)$$

$$= \frac{1}{\beta} + \ldots \cong T + O(T^{-1}), K = 1$$

$$\langle H^F \rangle = -\frac{\omega}{2} \left(\frac{\beta\omega}{2} \right) \frac{1}{\left(\frac{\beta\omega}{2} \right) \cot h \left(\frac{\beta\omega}{2} \right)} = -\frac{\omega^2}{4} \beta \left(\frac{1}{1 + \frac{1}{3} \left(\frac{\beta\omega}{2} \right)^2 \ldots} \right)$$

$$\cong -\frac{\omega^2}{4} \frac{1}{T}$$

Almost all energy is stored in the bosonic component.

Chapter 11
The Zeta Function in Quantum Electrodynamics (QED)

In this section the concept of the ζ-function regularization will be applied to computing the one-loop effective Lagrangian of spinor QED. We begin with the definition of the Green's function of a Dirac particle in an external field, which is described by a potential $A^\mu(x)$:

$$\left[\gamma \cdot \left(\frac{1}{i} \partial - eA \right) + m \right] G(x, x'|A) = \delta(x - x').$$

Hence the Green operator $G_+[A]$ is given by

$$(\gamma \Pi + m) G_+[A] = 1, \qquad \Pi_\mu = p_\mu - eA_\mu.$$

We are interested in evaluating the effective action $W^{(1)}[A]$ which is related to the one-loop effective Lagrangian $\mathcal{L}^{(1)}$ by

$$iW^{(1)}[A] = i \int d^3x dt \mathcal{L}^{(1)}. \tag{11.1}$$

So, on our way to computing $\mathcal{L}^{(1)}$ we have to determine the action $W^{(1)}$ first. This can be achieved by making use of the relation between $W^{(1)}[A]$ and $G_+[A]$,

$$iW^{(1)}[A] = -Tr \ln G_+[A], \tag{11.2}$$

or, equivalently:

$$iW^{(1)}[A] = - \ln \det G_+[A], \tag{11.3}$$

so that one now must calculate the determinant of the propagator $G_+[A]$. It is here where Riemann's ζ function allows us to solve the problem. In short, we have to evaluate

© The Author(s), under exclusive license to Springer Nature Switzerland AG 2021
W. Dittrich, *Reassessing Riemann's Paper*,
SpringerBriefs in History of Science and Technology,
https://doi.org/10.1007/978-3-030-61049-4_11

$$W^{(1)}[A] = i\zeta_2'(0) \, , \qquad\qquad (11.4)$$

which will be explained in extensive detail in the following pages.

The first step is to bring $W^{(1)}[A]$ into a form which leads to differential operators of the second order, allowing a simple calculation of the associated ζ function later on; first, we have from (11.2) and (11.3):

$$W^{(1)}[A] = iTr \ln G_+[A]$$
$$= -i \ln \det(m + \Pi\!\!\!/ - i\epsilon), \qquad \Pi\!\!\!/ = \gamma\Pi \, . \qquad (11.5)$$

Now we use

$$-\Pi\!\!\!/^{\,2} = \Pi^2 - \frac{e}{2}\sigma_{\mu\nu}F^{\mu\nu} = \Pi^2 - eB\sigma^3$$

and get

$$Tr \ln(m + \Pi\!\!\!/) + Tr \ln(m - \Pi\!\!\!/)$$
$$= Tr \ln(m^2 - \Pi\!\!\!/^{\,2})$$
$$= Tr \ln\left[(m^2 + \Pi^2)I_4 - eB\sigma^3\right] \, . \qquad (11.6)$$

If we denote the trace in spinor (coordinate) space as $tr_\gamma (tr_x)$ and the i-th eigenvalue of the matrix A as $EW_i(A)$, then

$$Tr \ln\left[(m^2 + \Pi^2)I_4 - eB\sigma^3\right]$$
$$\equiv tr_x tr_\gamma \ln\left[(m^2 + \Pi^2)I_4 - eB\sigma^3\right]$$
$$= tr_x \sum_{i=1}^{4} \ln EW_i \begin{bmatrix} m^2 + \Pi^2 - eB & & & 0 \\ & m^2 + \Pi^2 + eB & & \\ & & m^2 + \Pi^2 - eB & \\ 0 & & & m^2 + \Pi^2 + eB \end{bmatrix}$$
$$= 2tr_x \left[\ln(m^2 + \Pi^2 - eB) + \ln(m^2 + \Pi^2 + eB)\right] \, .$$

Substitution into (11.6) then gives

$$Tr \ln(m + \Pi\!\!\!/) + Tr \ln(m - \Pi\!\!\!/)$$
$$= 2tr_x \left[\ln(m^2 + \Pi^2 - eB) + \ln(m^2 + \Pi^2 + eB)\right]$$

or, with $\det_{\gamma x} \equiv \det$:

$$\ln \det{}_{\gamma x}(m + \Pi\!\!\!/) + \ln \det{}_{\gamma x}(m - \Pi\!\!\!/)$$
$$= 2 \ln \det{}_x(m^2 + \Pi^2 - eB) + 2 \ln \det{}_x(m^2 + \Pi^2 + eB) \, . \qquad (11.7)$$

But the determinant of $(m \pm \not\Pi)$ is a Lorentz scalar, which means in particular that it does not depend on the sign of the $\not\Pi$, i.e., it is

$$\det{}_{\gamma x}(m + \not\Pi) = \det{}_{\gamma x}(m - \not\Pi).$$

From (11.7) it follows that

$$\ln \det{}_{\gamma x}(m + \not\Pi) = \ln \det{}_x(m^2 + \Pi^2 + eB)$$
$$+ \ln \det{}_x(m^2 + \Pi^2 - eB), \tag{11.8}$$

which, put into (11.5), gives:

$$W^{(1)}[A] = -i\left[\ln \det{}_x(m^2 + \Pi^2 + eB) + \ln \det{}_x(m^2 + \Pi^2 + eB)\right]. \tag{11.9}$$

This equation is, however, not correct regarding the dimensions, since in (11.2) we left out the $G_+[0]$ in the denominator, the right side of (11.8) now has the dimension (mass)^2, while the left side in our units ($\hbar = c = 1$) is dimensionless. Thus, we introduce an arbitrary parameter μ with the dimension of a mass and replace (11.9) by

$$W^{(1)}[A] = i\left\{\ln \det{}_x[\mu^{-2}(m^2 + \Pi^2 + eB)]\right.$$
$$\left. + \ln \det{}_x[\mu^{-2}(m^2 + \Pi^2 + eB)]\right\}. \tag{11.10}$$

With the determinant definition (8.3) $\dot{W}^{(1)}[A]$, becomes

$$W^{(1)}[A] = -i\left\{\ln \exp\left[-\zeta'_{\mu^{-2}(m^2+\Pi^2+eB)}(0)\right]\right.$$
$$+ \ln \exp\left[-\zeta'_{\mu^{-2}(m^2+\Pi^2+eB)}(0)\right]\right\}$$
$$= i\zeta'_2(0) \tag{11.11}$$

with

$$\zeta_2(s) := \zeta_{\mu^{-2}(m^2-i\epsilon+\Pi^2+eB)}(s) + \zeta_{\mu^{-2}(m^2-i\epsilon+\Pi^2-eB)} \tag{11.12}$$

where we again substituted $m^2 - i\epsilon$ for m^2. Now we perform a Wick rotation by the substitution $t \to \tau = it$ and get for (11.12)

$$\zeta_2(s) = \zeta_{\mu^{-2}(m^2+\Pi_E^2+eB)}(s) + \zeta_{\mu^{-2}(m^2+\Pi_E^2-eB)}(s), \tag{11.13}$$

with the Euclidean momentum Π_E. In order to calculate the zeta ζ function (11.12) or (11.13) we need the spectrum of the operator

$$M^2 := m^2 + \Pi^2$$

$$= m^2 + \left(\frac{1}{i}\partial - eA\right)^2$$

$$= m^2 + \left(\frac{1}{i}\partial - eA\right)^2_{\parallel} + \left(\frac{1}{i}\partial - eA\right)^2_{\perp}$$

$$= m^2 + \partial_t^2 - \partial_3^2 - (\vec{p} - e\vec{A})^2_{\perp} \tag{11.14}$$

or its Euclidean analogue

$$M_E^2 = m^2 - \partial_\tau^2 - \partial_3^2 - (\vec{p} - e\vec{A})^2_{\perp}. \tag{11.15}$$

The spectrum of the operator $(\vec{p} - e\vec{A})^2_{\perp}$ is known from one-particle quantum mechanics; there it is shown that a particle, moving in a constant magnetic field $\vec{B} = B\hat{z}$ and having no momentum in z-direction, is described by a Hamilton operator

$$H = \frac{1}{2m}(\vec{p} - e\vec{A})^2_{\perp},$$

having the eigenvalues

$$E_n = \frac{e}{m}B\left(n + \frac{1}{2}\right), \quad n \in \mathbb{N}.$$

Thus, the last term of the sum of (11.15) has the eigenvalues

$$2(eB)\left(n + \frac{1}{2}\right), \quad n \in \mathbb{N}. \tag{11.16}$$

If we imagine further that we enclose the field in a very large normalization volume $\Omega = L^4$ of Euclidean space-time, then we can approximate $(-\partial_\tau^2 - \partial_3^2)$ by plane waves with eigenvalues $(k_0^2 + k_3^2)$, $k_0, k_3 \in \mathbb{R}$, and density $(L/2\pi)^2$. All together one then has

$$\text{spectrum } [m^2 + \Pi_E^2] =$$
$$\{m^2 + k_0^2 + k_3^2 + (2n + 1)eB | k_0, k_3 \in \mathbb{R}, \quad n \in \mathbb{N}\}. \tag{11.17}$$

In continuum approximation, the sum in (8.1) can be replaced by integrals, leading to

$$\zeta_2(s) = \mu^{2s}\Omega \sum_{n=0}^{\infty} \frac{eB}{2\pi} \int_{-\infty}^{\infty} \frac{dk_0 dk_3}{(2\pi)^2} \left\{ \left[m^2 + k_0^2 + k_3^2 + \right. \right.$$

$$\left. \left. +2(n+1)eB \right]^{-s} + \left[m^2 + k_0^2 + k_3^2 + 2neB \right]^{-s} \right\}. \tag{11.18}$$

Now we shall first evaluate the k-integrations; to do so, we use the formula

$$\int_0^{\infty} dx\, x^{\mu-1}(1+x^2)^{\nu-1} = \frac{1}{2}B\left(\frac{\mu}{2}, 1-\nu-\frac{\mu}{2}\right)$$

$$Re\,\mu > 0, \qquad Re\left(\nu = \frac{\mu}{2}\right) < 1 \tag{11.19}$$

and evaluate (11.18) for those values of s for which the integrals exist. (B designates the Beta function or the Euler integral of the first kind.) Thereafter, ζ_2 can be analytically continued to a meromorphic function of the whole complex plane. First we consider two special cases of (11.19):

$$\int_{-\infty}^{\infty} dx(a^2 + x^2)^{-s} = (a^2)^{\frac{1}{2}-s} B\left(\frac{1}{2}, s-\frac{1}{2}\right); \tag{11.20}$$

furthermore,

$$\int_{-\infty}^{\infty} dx(a^2 + x^2)^{\frac{1}{2}-s} = (a^2)^{1-s} B\left(\frac{1}{2}, s-1\right). \tag{11.21}$$

Applying (11.20) to (11.18) gives

$$\zeta_2(s) = \mu^{2s}\Omega B\left(\frac{1}{2}, s-\frac{1}{2}\right) \sum_{n=0}^{\infty} \frac{eB}{(2\pi)^3} \int_{-\infty}^{\infty} dk_3 \left\{ \left[m^2 + \right. \right.$$

$$\left. \left. +k_3^2 + (n+1)2eB \right]^{\frac{1}{2}-s} + \left[m^2 + k_3^2 + 2neB \right]^{\frac{1}{2}-s} \right\};$$

with (11.21) it follows that

$$\zeta_2(s) = \mu^{2s}\Omega B\left(\frac{1}{2}, s-\frac{1}{2}\right) B\left(\frac{1}{2}, s-1\right) \frac{eB}{(2\pi)^3}$$

$$\cdot \sum_{n=0}^{\infty} \left\{ \left[m^2 + 2(n+1)eB \right]^{1-2} + \left[m^2 + 2neB \right]^{1-s} \right\}. \tag{11.22}$$

If one uses the two functional equations of the Beta function

$$B(x, x) = 2^{1-2x} B\left(\frac{1}{2}, x\right)$$

$$B(x, x) B\left(x + \frac{1}{2}, x + \frac{1}{2}\right) = \pi \left[2^{4x-1} x\right]^{-1}, \tag{11.23}$$

one gets

$$B\left(\frac{1}{2}, s - \frac{1}{2}\right) B\left(\frac{1}{2}, s - 1\right)$$

$$= 2^{2s-2} 2^{2s-3} B\left(s - \frac{1}{2}, s - \frac{1}{2}\right) B(s - 1, s - 1)$$

$$= 2^{4s-5} \pi \left[2^{4(s-1)-1} (s - 1)\right]^{-1}$$

$$= \frac{\pi}{s - 1}.$$

Substitution into (11.22) gives

$$\zeta_2(s) = \mu^{2s} \Omega \frac{eB}{(2\pi)^3} \frac{\pi}{s - 1} \left\{ \sum_{n=0}^{\infty} \left[m^2 + 2eB(n + 1)\right]^{1-s} + \sum_{n=0}^{\infty} \left[m^2 + 2eBn\right]^{1-s} \right\}$$

$$= \mu^{2s} \Omega \frac{eB}{(2\pi)^3} \frac{\pi}{s - 1} (2eB)^{1-s}$$

$$\cdot \left\{ \sum_{n=0}^{\infty} \left(\frac{m^2}{2eB} + n + 1\right)^{1-s} + \sum_{n=0}^{\infty} \left(\frac{m^2}{2eB} + n\right)^{1-s} \right\}. \tag{11.24}$$

If one keeps in mind that the Riemann Zeta function in two arguments has the representation

$$\zeta(z, q) = \sum_{n=0}^{\infty} (q + n)^{-z}, \qquad \operatorname{Re} z > 1, \tag{11.25}$$

then it is clear that for the second sum one gets

$$\sum_{n=0}^{\infty} \left(\frac{m^2}{2eB} + n\right)^{1-s} = \zeta\left(s - 1, \frac{m^2}{2eB}\right)$$

and for the first

$$\sum_{n=0}^{\infty} \left(\frac{m^2}{2eB} + 1 + n\right)^{1-s} = \sum_{\nu=1}^{\infty} \left(\frac{m^2}{2eB} + \nu\right)^{1-s}$$

$$= \zeta\left(s - 1, \frac{m^2}{2eB}\right) - \left(\frac{m^2}{2eB}\right)^{1-s}.$$

Our end result for ζ_2 thus reads

$$\zeta_2(s) = \Omega\frac{eB}{8\pi^2}(s-1)^{-1}(2eB)^{1-s}\left[2\zeta(s-1,q)-q^{1-s}\right]\mu^{2s}, \qquad (11.26)$$

where we set

$$q := \frac{m^2}{2eB}. \qquad (11.27)$$

The derivative of this equation is

$$\begin{aligned}
\zeta'_2(s) = \Omega\frac{eB}{8\pi^2}\Big\{&-(s-1)^{-2}(2eB)^{1-s}\left[2\zeta(s-1,q)-q^{1-s}\right]\mu^{2s}\\
&-(s-1)^{-1}\ln(2eB)(2eB)^{1-s}\left[2\zeta(s-1,q)-q^{1-s}\right]\mu^{2s}\\
&+(s-1)^{-1}(2eB)^{1-s}\left[2\zeta'(s-1,q)+\ln q\cdot q^{1-s}\right]\mu^{2s}\\
&+(s-1)^{-1}(2eB)^{1-s}\left[2\zeta(s-1,q)-q^{1-s}\right]\ln\mu^2\mu^{2s}\Big\}
\end{aligned}$$

or, for $s = 0$:

$$\begin{aligned}
\zeta'_2(0) &= \Omega\frac{(eB)^2}{4\pi^2}\Big\{-[2\zeta(-1,q)-q]+\ln(2eB)[2\zeta(-1,q)-q]\\
&\quad-[2\zeta'(-1,q)+q\ln q]-\ln\mu^2[2\zeta(-1,q)-q]\Big\}\\
&= \Omega\frac{(eB)^2}{4\pi^2}\Big\{[q-2\zeta(-1,q)](1-\ln(2eB)+\ln\mu^2)\\
&\quad-[2\zeta'(-1,q)+q\ln q]\Big\}. \qquad (11.28)
\end{aligned}$$

To calculate $\zeta(-1, q)$ we need the following property of the ζ function:

$$\zeta(-n,q) = -\frac{B'_{n+2}(q)}{(n+1)(n+2)}, \qquad n \in \mathbb{N}. \qquad (11.29)$$

Here, B_n denotes the Bernoulli polynomials. In particular,

$$B_3(q) = q^3 - \frac{3}{2}q^2 + \frac{1}{2}q,$$

so that one gets

$$\begin{aligned}
\zeta(-1,q) &= -\frac{1}{6}B'_3(q)\\
&= -\frac{1}{2}\left(q^2-q+\frac{1}{6}\right). \qquad (11.30)
\end{aligned}$$

From (11.28) together with (11.27) it thus follows that

$$\zeta_2'(0) = \Omega \frac{(eB)^2}{4\pi^2} \left\{ \left[\left(\frac{m^2}{2eB} \right)^2 + \frac{1}{6} \right] \left(1 - \ln \frac{2eB}{\mu^2} \right) - \frac{m^2}{2eB} \ln \frac{m^2}{2eB} \right.$$

$$\left. -2\zeta' \left(-1, \frac{m^2}{2eB} \right) \right\} . \tag{11.31}$$

From this, according to (11.11) one can calculate via

$$W^{(1)}[A] = i\zeta_2'(0)$$

the effective action, and with

$$W^{(1)}[A] = \int d^3x dt \mathcal{L}^{(1)}$$

$$= \frac{1}{i} \int d^3x d\tau \mathcal{L}^{(1)}$$

$$= (-i)\Omega \mathcal{L}^{(1)} ,$$

the effective Lagrangian:

$$\mathcal{L}^{(1)}(B) = -\Omega^{-1} \zeta_2'(0)$$

$$= -\frac{(eB)^2}{4\pi^2} \left\{ \left(\frac{m^2}{2eB} \right)^2 + \frac{1}{6} - \left(\frac{m^2}{2eB} \right)^2 \ln \frac{2eB}{\mu^2} \right.$$

$$\left. -\frac{1}{6} \ln \frac{2eB}{\mu^2} - \frac{m^2}{2eB} \ln \frac{m^2}{2eB} - 2\zeta' \left(-1, \frac{m^2}{2eB} \right) \right\} . \tag{11.32}$$

This expression for $\mathcal{L}^{(1)}$ still depends on the arbitrary parameter μ, which can be fixed by additional conditions for $\mathcal{L}^{(1)}$. In the case of massive QED studied here, $\mu = m$ can be chosen as reference mass. With this choice (11.32) can be put—except for a constant—in the form

$$\mathcal{L}^{(1)}(B) = -\frac{1}{32\pi^2} \left\{ \left(2m^4 - 4m^2(eB) + \frac{4}{3}(eB)^2 \right) \right.$$

$$\cdot \left[1 + \ln \frac{m^2}{2eB} \right] + 4m^2(eB)$$

$$\left. -3m^4 - 16(eB)^2 \zeta' \left(-1, \frac{m^2}{2eB} \right) \right\} , \tag{11.33}$$

which can be proven by simply multiplying out Eq. (11.33). One can very easily specialize the above calculation to the case of vanishing Fermion mass ($m = 0$). For this special case, one reads from (11.24)

$$\zeta_2(s) = \mu^{2s}\Omega\frac{(eB)}{(2\pi)^3}\frac{\pi}{s-1}(2eB)^{1-s}\left\{\sum_{n=0}^{\infty}(n+1)^{1-s} + \sum_{n=0}^{\infty}n^{1-s}\right\}$$

$$= \mu^{2s}\Omega\frac{(eB)}{(2\pi)^2}\frac{1}{s-1}(2eB)^{1-s}\zeta(s-1). \tag{11.34}$$

Here, in the last line, the representation

$$\zeta(z) = \sum_{n=1}^{\infty}n^{-z}, \qquad \mathrm{Re}\,z > 1$$

of the normal Riemann Zeta function was used. Differentiation of (11.34) gives

$$\zeta'_2(s) = \Omega\frac{eB}{(2\pi)^2}\Big[-(s-1)^{-2}(2eB)^{1-s}\zeta(s-1)\mu^{2s}$$

$$+ (s-1)^{-1}[-\ln 2eB](2eB)^{1-s}\zeta(s-1)\mu^{2s}$$

$$+ (s-1)^{-1}(2eB)^{1-s}\zeta'(s-1)\mu^{2s}$$

$$+ (s-1)^{-1}(2eB)^{1-s}\zeta(s-1)\ln\mu^2 \cdot \mu^{2s}\Big].$$

Now, we set $s = 0$ again:

$$\zeta'_2(0) = \Omega\frac{eB}{4\pi^2}\Big[-(2eB)\zeta(-1) + (2eB)\zeta(-1)\ln(2eB)$$

$$-(2eB)\zeta'(-1) - (2eB)\zeta(-1)\ln\mu^2\Big].$$

Since for the Riemann Zeta function

$$\zeta(z, 1) = \zeta(z)$$

is valid ($z\epsilon\mathbb{C}$ arbitrary), then

$$\zeta(-1) = \zeta(-1, q = 1) = -\frac{1}{12} \; !$$

And thus

$$\zeta'_2(0) = -\Omega\frac{(eB)^2}{24\pi^2}\left[\ln\frac{2eB}{\mu^2} - 1 + 12\zeta'(-1)\right].$$

Since

$$\mathcal{L}^{(1)} = -\Omega^{-1}\zeta'_2(0),$$

the one-loop effective Lagrangian for QED with massless Fermions is then given by

$$\mathcal{L}_{m=0}^{(1)}(B) = \frac{(eB)^2}{24\pi^2}\left[\ln\frac{2eB}{\mu^2} - 1 + 12\zeta'(-1)\right]$$

$$= \frac{\alpha B^2}{6\pi}\left[\ln\frac{2eB}{\mu^2} - 1 + 12\zeta'(-1)\right]. \tag{11.35}$$

If we would replace the undetermined parameter μ in the denominator of the logarithm by the electron mass m, then this would be precisely the expression for the Lagrangian of the massive theory in the limiting case of strong fields! But since there exists no natural mass scale in the theory with $m = 0$ with respect to which one can measure the field strengths, it is not possible with the help of the renormalization conditions applied to $\mathcal{L}$ to eliminate μ from $\mathcal{L}_{m=0}^{(1)}$. Nevertheless, the physical content of (11.35) cannot depend on μ, i.e.,

$$\mu\frac{d}{d\mu}\mathcal{L}(\mu, \alpha(\mu), B(\mu)) = 0 \tag{11.36}$$

must be valid: here we have indicated with the arguments that $\mathcal{L}$ also has an implicit dependence on μ via α and B, since with the choice of a particular value for μ, one also decides on a special renormalized coupling constant α or field strength which, as we have seen in the case of them massive theory, exactly coincides with the physical coupling constants or field strengths for $\mu = m$. With (11.36) we are led to the so-called renormalization group equation

$$\left[\mu\frac{\partial}{\partial\mu} + \mu\frac{d\alpha}{d\mu}\frac{\partial}{\partial\alpha} + \mu\frac{dB}{d\mu}\frac{\partial}{\partial B}\right]\mathcal{L}(\mu, \alpha, B) = 0. \tag{11.37}$$

This partial differential equation tells us that an infinitesimal change in μ can be compensated for by a corresponding change in α and a rescaling of B. The above equation can be further simplified by utilizing the relationship between the renormalized quantities e, B and the bare e_0, B_0:

$$e = e_0 Z_3^{1/2}$$
$$B = B_0 Z_3^{-1/2}. \tag{11.38}$$

Furthermore, $\alpha = Z_3\alpha_0$ and $eB = e_0 B_0$, so that for the derivatives required in (11.37) we get

$$\frac{d\alpha}{d\mu} = \alpha_0\frac{dZ_3}{d\mu} = \frac{\alpha}{Z_3}\frac{dZ_3}{d\mu} = \alpha\frac{d\ln Z_3}{d\mu}$$

and

$$\frac{dB}{d\mu} = B_0\frac{dZ_3^{-1/2}}{d\mu} = -\frac{1}{2}B_0 Z_3^{-3/2}\frac{dZ_3}{d\mu}$$

$$= -\frac{1}{2}B\frac{d\ln Z_3}{d\mu}.$$

Here we used the fact that the bare quantities e_0 and B_0 are naturally independent of the parameter μ, which is only introduced during the regularization process. Now we define

$$\beta_\zeta(\alpha) := \mu \frac{d \ln Z_3}{d\mu}.$$ (11.39)

and get

$$\mu \frac{d\alpha}{d\mu} = \alpha \beta_\zeta(\alpha)$$

$$\mu \frac{dB}{d\mu} = -\frac{1}{2}\beta_\zeta(\alpha)B,$$

which, together with (11.37) gives

$$\left[\mu \frac{\partial}{\partial\mu} + \alpha\beta_\zeta(\alpha)\frac{\partial}{\partial\alpha} - \frac{1}{2}\beta_\zeta(\alpha)B\frac{\partial}{\partial B} \right] \mathcal{L}(\mu,\alpha,B) = 0.$$ (11.40)

Note that this equation contains only renormalized quantities. By now substituting the known one-loop approximation for $\mathcal{L}$, we can directly calculate the function β_ζ. From

$$\mathcal{L}(\mu,\alpha,B) = -\frac{1}{2}B^2 + \frac{\alpha B^2}{6\pi} \ln \frac{2\sqrt{4\pi\alpha}B}{\mu^2} + \frac{\alpha B^2}{6\pi}C$$

with

$$C = -1 + 12\zeta'(-1),$$

it follows that

$$\left[\mu\frac{\partial}{\partial\mu} + \alpha\beta_\zeta(\alpha)\frac{\partial}{\partial\alpha} - \frac{1}{2}\beta_\zeta B\frac{\partial}{\partial B} \right]\mathcal{L}(\mu,\alpha,B)$$

$$= -\frac{\alpha B^2}{3\pi}$$

$$+ \beta_\zeta(\alpha)\frac{\alpha B^2}{6\pi}\left[\ln\frac{2eB}{\mu^2} + C \right] + \beta_\zeta(\alpha)\frac{\alpha B^2}{12\pi^2}$$

$$+ \frac{1}{2}\beta_\zeta(\alpha)B^2 - \beta_\zeta(\alpha)\frac{\alpha B^2}{6\pi}\left[\ln\frac{2eB}{\mu^2} + C \right]$$

$$- \beta_\zeta(\alpha)\frac{\alpha B^2}{12\pi^2}$$

$$= \left(\frac{1}{2}\beta_\zeta(\alpha) - \frac{\alpha}{3\pi} \right)B^2$$

$$\overset{(11.40)}{=} 0.$$

The validity of (11.36) implies that

$$\beta_\zeta(\alpha) = \frac{2}{3}\frac{\alpha}{\pi} \, .$$

(11.41)

Summary of Euler-Riemann Formulae

$$\zeta(s) = \prod_{p,\text{prime}} \frac{1}{1 - p^{-s}} = \sum_{n=1}^{\infty} \frac{1}{n^s}, \text{Re}(s) > 1. \tag{1}$$

Prime number counting function:

$$\Pi(x) = \sum_{p^n < x} \frac{1}{n} = \sum_{p} \sum_{n=1}^{\infty} \frac{1}{n} \Theta(x - p^n). \tag{2}$$

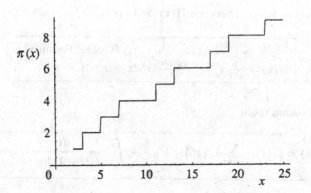

$$\pi(x) = \sum_{p \leq x} 1. \tag{3}$$

$\pi(x)$ = number of primes less than or equal to x.

Primes under 25 are 2, 3, 5, 7, 11, 13, 17, 19 and 23, i.e., $\pi(3) = 2$, $\pi(10) = 4$, ... $\pi(25) = 9$.

$\pi(x)$ is locally irregular.

Let x be a positive real number.

How many primes are there less than x?

The answer was given by B. Riemann in an 1859 paper entitled, "On the Number of Primes Less than a Given Magnitude."

Relation between $\Pi(x)$ and $\pi(x)$; Riemann's formula:

$$\Pi(x) = \sum_{n=1}^{\infty} \frac{1}{n} \pi(x^{\frac{1}{n}}). \qquad (4)$$

Inversion of (4):

$$\pi(x) = \sum_{n=1}^{\infty} \frac{\mu(n)}{n} \Pi(x^{\frac{1}{n}}) = \Pi(x) - \frac{1}{2}\Pi(x^{\frac{1}{2}}) - \frac{1}{3}\Pi(x^{\frac{1}{3}}) - \cdots \qquad (5)$$

The Möbius function is defined via inversion of the ζ function:

$$\prod_p \left(1 - \frac{1}{p^s}\right) = \frac{1}{\zeta(s)} = \sum_{n=1}^{\infty} \frac{\mu(n)}{n^s} = 1 - \frac{1}{2^s} - \frac{1}{3^s} - \frac{1}{5^s} + \frac{1}{6^s} - \frac{1}{7^s} + \cdots$$

with $\mu(1) = 1$, $\mu(2) = -1$, $\mu(3) = -1$, $\mu(4) = 0$, $\mu(5) = -1$, $\mu(6) = 1$, $\mu(7) = -1$, ...

Very important relation between $\Pi(x)$ and $\zeta(s)$:

$$\boxed{\begin{array}{l} \frac{\log \zeta(s)}{s} = \int_0^{\infty} \Pi(x) x^{-s-1} dx \dots \dots \text{"Golden Formula"} \\ \Pi(x) = \frac{1}{2\pi i} \int_{a-i\infty}^{a+i\infty} \frac{\log(\zeta(s))}{s} x^s ds, \quad a > 1. \end{array}} \qquad (6)$$

Riemann's main result:

$$\boxed{\Pi(x) = \mathrm{L_i}(x) - \sum_{\varrho} \mathrm{L_i}(x^{\varrho}) + \log\frac{1}{2} + \int_x^{\infty} \frac{dt}{t(t^2 - 1)\log t}, x > 1.} \qquad (7)$$

Differentiation of $\Pi(x)$:

$$\frac{d\Pi(x)}{dx} = \left(\frac{1}{\log x} - \sum \frac{x^{\varrho-1}}{\log x} - \frac{1}{x(x^2 - 1)\log x}\right), x > 1. \qquad (8)$$

When we write

$$d\Pi = \left(\frac{d\Pi(x)}{dx}\right)dx, \quad \Pi(x) = \sum_{p}\sum_{n=1}^{\infty}\frac{1}{n}\Theta(x - p^n), \tag{9}$$

we see that the measure $d\Pi$ is dx times the density of primes plus ½ the density of prime squares, plus the density of prime cubes, etc.

Thus $\frac{1}{\log x}$ alone should not be considered an approximation only to the density of primes as Gauss suggested, but rather to $\frac{d\Pi(x)}{dx}$, i.e., to the density of primes plus ½ the density of prime squares, plus etc.

Gauss never published anything about the prime number theorem. But he knew from looking at the prime number tables up to two million that for large values of x that

$$\pi(x) \sim \frac{x}{\log x}, \quad x \to \infty; \quad \pi(x) \sim Li(x) = \int_{2}^{x}\frac{dx}{\log x}. \tag{10}$$

This result was found by the 15-year-old Gauss around 1792/93. It was communicated in a letter to his friend J.F. Encke on December 24, 1849, when Gauss was 72 years old.

Ramanujan proved the following formula:

$$\frac{d\pi(x)}{dx} = \frac{1}{x\log x}\sum_{n=1}^{\infty}\frac{\mu(n)}{n}x^{\frac{1}{n}}. \tag{11}$$

Evidently there are no zeros of the ζ function involved. At least one can learn from expanding

$$\frac{d\pi(x)}{dx} = \frac{1}{x\log x}\left(x - \frac{1}{2}x^{\frac{1}{2}} - \frac{1}{3}x^{\frac{1}{3}} - \frac{1}{5}x^{\frac{1}{5}} - \cdots\right)$$

$$= \frac{1}{\log x} - \frac{1}{2}\frac{1}{x^{\frac{1}{2}}\log x} - \cdots \tag{12}$$

that the density of primes decreases rapidly as x takes on larger and larger values.

However, there exists an equivalent to Riemann's $\Pi(x)$ given by v. Mangoldt's formula for $\psi(x)$. This function was introduced by Chebyshev (1821–1894). He added the logs of prime numbers, i.e., he wrote

$$\psi(x) = \sum_{p^n \leq x} \log p, \tag{13}$$

e.g.,

$$\psi(20) = (\log 2 + \log 3 + \log 5 + \log 7 + \log 11 + \log 13 + \log 17$$
$$+ \log 19) + (\log 2 + \log 3) + (\log 2) + (\log 2) = 19.2656$$

where within the brackets, we have $p < x^{\frac{1}{n}}, n = 1, 2, 3 \dots$.

For this function, v. Mangoldt found a most remarkable expression:

$$\psi(x) = x - \log(2\pi) - \frac{1}{2}\log\left(1 - \frac{1}{x^2}\right) - \sum_{\zeta(\varrho)=0} \frac{x^\varrho}{\varrho}. \qquad (14)$$

Because of the simplicity of this formula, it is meanwhile considered preferable to that of $\Pi(x)$.

The following graphs show the Chebyshev step function and its successive approximations by the von Mangoldt function with increasing zeros of the ζ function. The steps occur at prime numbers and at numbers which are powers of prime numbers, e.g., at x = 13, 17, 19, but also at x = 8 = 2^3, 25 = 5^2, 27 = 3^3, etc.

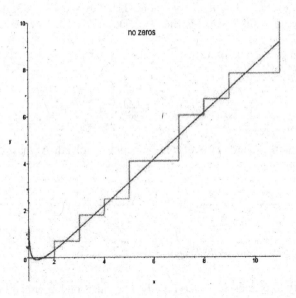

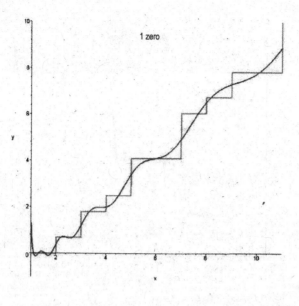

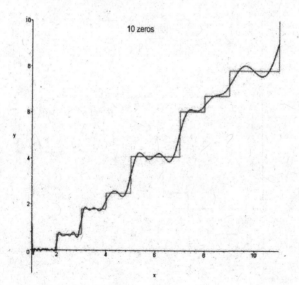

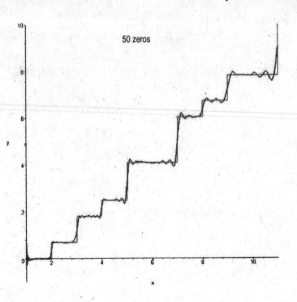

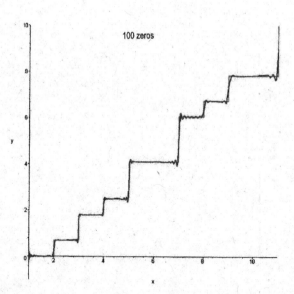

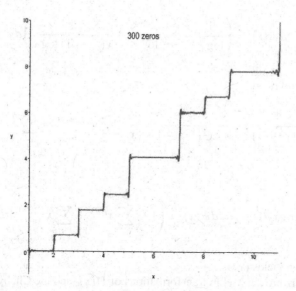

300 zeros

Let us try to understand Eq. (14) by beginning with Riemann's "Golden Formula" (6):

$$log\zeta(s) = \int_0^\infty \Pi(x)sx^{-s-1}dx = -s\int_0^\infty dxd\Pi(x)\frac{x^{-s}}{-s}$$

$$= \int_0^\infty x^{-s}d\Pi(x). \text{ (Stieltjes integral, unknown to Riemann?)} \qquad (15)$$

Differentiation leads to

$$-\frac{\zeta'(s)}{\zeta(s)} = \int_0^\infty x^{-s}logx\, d\Pi(x) = \int_0^\infty x^{-s}d\psi(x)$$

$$= -\int_0^\infty -sx^{-s-1}\psi(x)dx \qquad (16)$$

$$d\psi(x) = logx\, d\Pi(x) = s\int_1^\infty \psi(x)x^{-s-1}dx. \qquad (17)$$

From (14) we take $\frac{1}{2}log(1-\frac{1}{x^2}) = \sum_n \frac{1}{2n}x^{-2n}$ and differentiate $\frac{d}{dx}\sum_n \frac{1}{2n}x^{-2n} = -\sum_n x^{-2n-1}$.

Hence we obtain from (14)

$$\frac{d\psi}{dx} = \left(1 + \sum_n x^{-2n-1} - \sum x^{\varrho-1}\right),$$

and from (8):

$$d\Pi(x) = \left(\frac{1}{\log x} - \sum \frac{x^{\varrho-1}}{\log x} - \frac{1}{x(x^2-1)\log x} \right) dx.$$

With the aid of

$$\frac{1}{x(x^2-1)} = \frac{1}{x \cdot x^2 \left(1 - \frac{1}{x^2}\right)} = \frac{1}{x^3} \sum_{1}^{\infty} x^{-2(n-1)} = \sum_{n} x^{-2n-1}$$

we obtain

$$log x\, d\Pi(x) = d\psi(x) = \left(1 - \sum x^{\varrho-1} + \sum_{n} x^{-2n-1} \right) dx.$$

Integration yields (14).

Here, then, are the Mellin transformations of $\Pi(x)$ and the Chebyshev function $\psi(x)$:

$$\frac{\log \zeta(s)}{s} = \int_{0}^{\infty} \Pi(x) x^{-s-1} dx, \ Re(s) > 1, {}_{Riemann\,1859} \tag{18}$$

$$-\frac{1}{s} \frac{\zeta'(s)}{\zeta(s)} = \int_{1}^{\infty} \psi(x) x^{-s-1} dx, \ Re(s) > 1, v._{Mangoldt\,1895} \tag{19}$$

The inverse of these formulae is given by

$$\Pi(x) = \frac{1}{2\pi i} \int_{a-i\infty}^{a+i\infty} \frac{\log(\zeta(s))}{s} x^s ds, \ a > 1. \tag{20}$$

$$\psi(x) = \frac{1}{2\pi i} \int_{a-i\infty}^{a+i\infty} -\frac{1}{s} \frac{\zeta'(s)}{\zeta(s)} \frac{x^s}{s} ds. \tag{21}$$

The right-hand side can be worked out and leads to Riemann's and v. Mangold's explicit expressions,

$$\Pi(x) = Li(x) - \sum_{Im\varrho>0} \left(Li(x^\varrho) + Li(x^{1-\varrho}) \right) - log2 + \int_{x}^{\infty} \frac{dt}{t(t^2-1)\log t} \tag{22}$$

$$\psi(x) = x - \log(2\pi) - \frac{1}{2} \log\left(1 - \frac{1}{x^2}\right) - \sum_{\zeta(\varrho)=0} \frac{x^\varrho}{\varrho}. \tag{23}$$

The Riemann ζ function provides an exact formula for $\pi(x)$ by summing over all the non-trivial zeros of the ζ function (in order of increasing magnitude):

$$\pi(x) = \sum_{n=1}^{\infty} \frac{\mu(n)}{n} Li(x^{\frac{1}{n}}) - \sum_{n=1}^{\infty} \frac{\mu(n)}{n} \sum_{\varrho} Li\left(x^{\frac{\varrho}{n}}\right)$$

$$+ \sum_{n=1}^{\infty} \frac{\mu(n)}{n} \int_{x^{\frac{1}{n}}}^{\infty} \frac{dt}{t(t^2 - 1)\log t} - \log 2 \sum_{n=1}^{\infty} \frac{\mu(n)}{n}.$$

The graph of this formula approaches the value in the middle of each step. At the primes, the graph of $\pi(x)$ takes a step up one unit.

Without proof, some of Riemann's formulae from his paper, "On the Number of Primes Less than a Given Magnitude":

$$\zeta(s) = 2^s \pi^{s-1} \sin\left(\frac{s\pi}{2}\right) \Gamma(1-s)\zeta(1-s), \quad Re(s) \leq 0, s\ complex$$

or, equivalently:

$$\boxed{\zeta(1-s) = \frac{2}{(2\pi)^s} \cos\left(\frac{s\pi}{2}\right) \Gamma(s)\zeta(s)}$$

$$\zeta_R(-1) = -\frac{1}{12}!$$

Riemann's entire function $\xi(s)$:

$$\xi(s) = \frac{1}{2}s(s-1)\pi^{-\frac{s}{2}}\Gamma\left(\frac{s}{2}\right)\zeta(s)$$

$$\xi(s) = \xi(1-s)$$

Functional equation:

$$\Gamma\left(\frac{s}{2}\right)\pi^{-\frac{s}{2}}\zeta(s) = \Gamma\left(\frac{1-s}{2}\right)\pi^{-\frac{1}{2}(1-s)}\zeta(1-s)$$

We obtain the right-hand side from the left-hand side by replacing s by $(1-s)$!

$$\boxed{\zeta'(-2n) = (-1)^n \frac{(2n)!}{2(2\pi)^{2n}}\zeta(2n+1)} \quad n \in \mathbb{N}$$

Proof. Riemann's functional equation:

$$\zeta(1-s) = 2(2\pi)^{-s}\Gamma(s)\zeta(s)\cos\left(\frac{\pi}{2}s\right)$$

derivative:

$$-\zeta'(1-s) = \left(2(2\pi)^{-s}\Gamma(s)\zeta(s)\right)' \cos\left(\frac{\pi}{2}s\right)$$

$$+ 2(2\pi)^{-s}\Gamma(s)\zeta(s)\left(-\frac{\pi}{2}\sin\left(\frac{\pi}{2}s\right)\right).$$

$$s = 2n+1 : \cos\left(\frac{\pi}{2}(2n+1)\right) = 0, \sin\left(\frac{\pi}{2}(2n+1)\right) = (-1)^n\frac{\pi}{2}.$$

$$> \zeta'(-2n) = \pi(2\pi)^{-(2n+1)}\Gamma(2n+1)\zeta(2n+1)(-1)^n$$

$$\Gamma(n) = (n-1)! = (-1^n)\frac{(2n)!}{2(2\pi)^{2n}}\zeta(2n+1)$$

$$\Gamma(2n+1) = (2n)!$$

$$s = 2n \qquad -\zeta'(1-2n) = \left(2(2\pi)^{-s}\Gamma(s)\zeta(s)\right)'|_{s=2n}\cos(\pi n)$$

$(1-2n)$odd and integer

$$\sin(n\pi) = 0$$

$$\cos(n\pi) = (-1)^n \quad \zeta'(1-2n) = \left((-1)^{n+1}2\right)\left((2\pi)^{-s}\Gamma(s)\zeta(s)\right)'|_{s=2n}$$

$$\left((2\pi)^{-s}\Gamma(s)\zeta(s)\right)'_{s=2n} = \left((2\pi)^{-s}\right)'\Gamma(s)\zeta(s) + (2\pi)^{-s}\left(\Gamma(s)\zeta(s)\right)'|_{s=2n}$$

$$\frac{d}{ds}e^{-s\log 2\pi} = -\log(2\pi)(2\pi)^{-s}$$

$$s = 2n: -\log(2\pi) \cdot (2\pi)^{-2n}$$

$$= -\frac{1}{(2\pi)^{2n}}\log(2\pi)\Gamma(2n)\zeta(2n)$$

$$+ \frac{1}{(2\pi)^{2n}}\left[\Gamma'(s)\zeta(s) + \Gamma(s)\zeta'(s)\right]_{s=2n}$$

$$= \frac{1}{(2\pi)^{2n}}\Delta$$

$$\Delta = \left(\Gamma(2n)\psi(2n)\zeta(2n) + \Gamma(2n)\zeta'(2n)\right)$$

$\Gamma'(s) = \Gamma(s)\psi(s)$

$\psi(s)$ *digamma function*

$> \zeta'(1-2n) = 2(-1)^{n+1}\frac{\Gamma(2n)}{(2\pi)^{2n}}\left[-\log(2\pi)\zeta(2n) + \psi(2n)\zeta(2n) + \zeta'(2n)\right]$

$$\boxed{> \zeta'(1-2n) = (-1)^{n+1}\frac{2\Gamma(2n)}{(2\pi)^{2n}}\left[(-\log(2\pi) + \psi(2n))\zeta(2n) + \zeta'(2n)\right]}$$

Claim: $\zeta'(-1)$ may be written as a function of $\zeta(2)$ and $\zeta'(2)$!

Relation to Glaisher-Kinkelin constant A:

$$\boxed{\zeta'(-1) = \frac{1}{12} - \log A}$$

Start from $\zeta'(1 - 2n)|_{n=1} = \zeta'(-1) = \frac{1}{12} - \log A$

$$= \frac{2\Gamma(2)}{(2\pi)^2}\left((-\log(2\pi) + \psi(2))\zeta(2) + \zeta'(2)\right)$$

$\Gamma(2) = 1:$ $\zeta'(-1) = \frac{1}{2\pi^2}\left((-\log 2\pi + 1 - \gamma)\zeta(2) + \zeta'(2)\right)$

$\psi(2) = 1 - \gamma$

$\ln\gamma = C, Euler's const.$ $> 1 - 12\log A = 12 \cdot \zeta'(-1)$

$$= 12 \cdot \frac{1}{2\pi^2}\left((-\log 2\pi + 1 - \gamma)\zeta(2) + \zeta'(2)\right)$$

$$= \frac{6}{\pi^2}\left((-\log 2\pi + 1 - \gamma)\zeta(2) + \zeta'(2)\right)$$

$$\zeta(2) = \frac{\pi^2}{6} = \frac{1}{\zeta(2)}\left((-\log 2\pi + 1 - \gamma)\zeta(2) + \zeta'(2)\right)$$

$$1 - 12\log A = (-\log(2\pi) + 1 - \gamma) + \frac{\zeta'(2)}{\zeta(2)}$$

$$> \frac{\zeta'(2)}{\zeta(2)} = 1 - 12\log A + \log(2\pi) - 1 + \gamma$$

$$> \zeta'(2) = \zeta(2)(-12\log A + \log(2\pi) + \gamma)$$

$$\boxed{\zeta'(2) = \frac{1}{6}\pi^2(-12\log A + \log(2\pi) + \gamma)}$$

$$= 0.93754825\ldots$$

$\zeta'(2) = \zeta(2)\left[-12\log A + \log(2\pi) + \gamma\right]$ $\zeta(2) = \frac{\pi^2}{6}, \gamma\, Eul.const.$

$$\gamma = 0.57721566490\ldots$$

$$A = Glaischer - Kinkelin const.$$

$$sometimes\, C = \log\gamma = Eul.const$$

$$> \frac{\zeta'(2)}{\zeta(2)} = -12\log A + \log(2\pi) + \gamma$$

$$> 12\log A = \log(2\pi) + \gamma - \frac{\zeta'(2)}{\zeta(2)}$$

$$> \log A = \frac{1}{12}\left[\log(2\pi) + \gamma - \frac{\zeta'(2)}{\zeta(2)}\right]$$

$$A = \exp\left(\frac{1}{12}\left[\log(2\pi) + \gamma - \frac{\zeta'(2)}{\zeta(2)}\right]\right)$$

$$A = (2\pi)^{\frac{1}{12}}\left[e^{\left(\frac{\pi^2}{6}\gamma - \zeta'(2)\right)}\right]^{\frac{1}{2\pi^2}}$$

$$log\, A =: L_1 = \frac{1}{12} - \zeta'(-1) = \frac{1}{12}\log(2\pi) + \frac{\gamma}{12} - \frac{1}{2\pi^2}\zeta'(2)$$

$$\zeta'(-1) = -0.165421\ldots, \zeta'(2) = -0.9375482543\ldots$$

$$L_1 = 0.248754477\ldots$$

! old definition of $C =: log\,\gamma = 0.5772\ldots$ *Euler*

Then $-1 + 12\zeta'(-1) = -\log(2\pi\gamma) + \frac{6}{\pi^2}\zeta'(2)$ follows from Riemann's functional equation for the ζ function.

$$\zeta'(0) = -\frac{1}{2}log(2\pi) \quad \zeta\left(\frac{1}{2}\right) = -1.46035450880\ldots$$

$$\zeta'(-2) = -\frac{\zeta(3)}{4\pi^2} \quad \zeta'\left(\frac{1}{2}\right) = \frac{1}{4}(6log2) + 2log\pi + \pi + 2\gamma\zeta\left(\frac{1}{2}\right)$$

$$\zeta'(-4) = -\frac{3\zeta(5)}{4\pi^4} \quad \zeta(-1) = -\frac{1}{12}$$

$$\zeta'(-6) = -\frac{45\zeta(7)}{8\pi^6} \quad \zeta'(-1) = \frac{1}{12} - log\, A \quad A = \text{Glaischer} - \text{Kinkelin constant}$$

$$\zeta'(-8) = -\frac{315\zeta(9)}{4\pi^8}$$

Riemann: Example:

$$\zeta'(-4) = \frac{\pi}{2}2^{-4}\pi^{-5}\cos(-2\pi)\Gamma(5)\zeta(5)$$

$$= \frac{\pi}{2}\cdot\frac{1}{16}\frac{1}{\pi^5}\cdot 1\cdot 24\zeta(5) = \frac{3}{4}\frac{1}{\pi^4}\zeta(5)$$

$$\zeta(2) = \frac{\pi^2}{6}, \zeta(3) = 1.2020569032, \quad \zeta(2) = 1.644934\ldots$$

$$\zeta(4) = \frac{\pi^4}{90}, \zeta(5) = 1.0369277551, \quad \zeta(4) = 1.0823232337\ldots$$

$$\zeta(6) = \frac{\pi^6}{945}, \zeta(7) = 1.0083492774, \quad \zeta(6) = 1.01734306\ldots$$

$$\zeta(8) = \frac{\pi^8}{9450}, \zeta(9) = 1.0020083928,$$

$$\zeta(10) = \frac{\pi^{10}}{93555}$$

$$\xi(2) = \frac{\zeta(2)}{\pi} = \frac{\pi}{6}, \xi(4) = \frac{6}{\pi^2}\zeta(4) = \frac{\pi^2}{15}$$

$$\xi(3) = \frac{3}{2\pi}\zeta(3), \xi(5) = \frac{15}{2\pi^2}\zeta(5)$$

$$\zeta(3) = 1 + \frac{1}{2^3} + \frac{1}{3^3} + \frac{1}{4^3} + \frac{1}{5^3} = 1 + \frac{1}{8} + \frac{1}{27} + \frac{1}{64} + \frac{1}{125} + \cdots$$

$$\zeta(3) \approx 1.202056903159\ldots$$

$\zeta(3)$ Apéry's constant; Apéry (14.9.1916 –18.12.1994) proved irrationality of $\zeta(3)$ in 1977 at the age of 61.

Apéry proved that the continued fraction of $\zeta(3)$ is never ending

$$\{1, 4, 1, 18, 1, 1, 1, \ldots\} \quad : |1 + \cfrac{1}{4 + \cfrac{1}{1 + \cfrac{1}{18 + \frac{1}{\cdots}}}}$$

Appendix A
Supplements and Appendix

A.1 Supplements

The Riemann ζ function can be extended meromorphically into the region $\{s : \Re(s) > 0\}$ in and on the right of the critical strip $\{s : 0 \leq \Re(s) < 1\}$. This is a sufficient region of meromorphic continuation for many applications in analytic number theory. The zeroes of the ζ function in the critical strip are known as the non-trivial zeroes of ζ.

It is remarkable that ζ obeys a functional equation establishing a symmetry across the critical line $\{s : \Re(s) = \frac{1}{2}\}$ rather than the real axis. One consequence of this symmetry is that the ζ function may be extended meromorphically to the entire complex plane with a simple pole at $s = 1$ and no other poles. For all $\mathbb{C} \setminus \Re(s) = 1$ including the strip we have the functional equation:

$$\zeta(s) = 2^s \pi^{s-1} \sin(\frac{s\pi}{2})\Gamma(1-s)\zeta(1-s), \quad \Re(s) < 0 \tag{A.1}$$

or, equivalently, the identity between meromorphic functions $\zeta(s)$:

$$\zeta(1-s) = \frac{2}{(2\pi)^s} \cos\left(\frac{s\pi}{2}\right)\Gamma(s)\,\zeta(s). \tag{A.2}$$

The analytical continuation given here allows one to connect $\zeta(s)$ for positive values of $\Re(s)$ with the same for negative values, for instance:

$$\zeta(-1) = 2^{-1}\pi^{-2}(-1)\Gamma(2)\zeta(2) = \frac{1}{2} \cdot \frac{1}{\pi^2} \cdot (-1) \cdot 1 \cdot \frac{\pi^2}{6} = -\frac{1}{12}, \tag{A.3}$$

i.e.,

$$\zeta_R(-1) = -\frac{1}{12}, \tag{A.4}$$

© The Author(s), under exclusive license to Springer Nature Switzerland AG 2021
W. Dittrich, *Reassessing Riemann's Paper*,
SpringerBriefs in History of Science and Technology,
https://doi.org/10.1007/978-3-030-61049-4

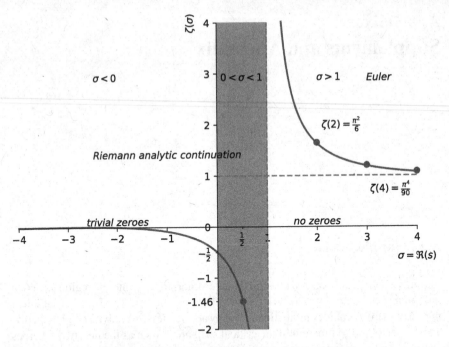

Fig. A.1 The different domains of definition of Riemann's ζ function of (A.5)

where the subscript R is added to distinguish Riemann's ζ from Euler's ζ, of which it is an extension, i.e., (Fig. A.1)

$$\zeta(x) = \sum_{n-1}^{\infty} \frac{1}{n^x} = \prod_{P \text{ prime}} \frac{1}{1 - p^{-x}} \text{ converging for } x > 1$$

$$\frac{1}{1^x} + \frac{1}{2^x} + \frac{1}{3^x} + \cdots = \prod_{P \text{ prime}} \frac{p^x}{p^x - 1} = \left(\frac{2^x}{2^x - 1}\right)\left(\frac{3^x}{3^x - 1}\right)\left(\frac{5^x}{5^x - 1}\right)\cdots$$

When we extend this function into the whole complex s plane, then Riemann's ζ function comes in three different representations:

$$\zeta(s) = \begin{cases} \sum_{n=1}^{\infty} \frac{1}{n^s} = \prod_p \text{ prime} \frac{p^s}{p^s - 1}, & \Re(s) > 1 \\ (1 - 2^{1-s}) \sum_{n=1}^{\infty} \frac{(-1)^{n+1}}{n^s}, & 0 < \Re(s) < 1 \\ 2^s \pi^{s-1} \sin(\frac{s\pi}{2})\Gamma(1 - s)\zeta(1 - s), & \Re(s) < 0 \end{cases} \quad (A.5)$$

where is $\zeta(s)$ equal to zero?

1. No zeroes for $\Re(s) > 1$ since here $\zeta(s) > 0$.
2. Non-trivial zeroes in the strip $0 < \Re(s) < 1$, symmetric around $\Re(s) = \frac{1}{2}$.
3. Trivial zeroes for s = $-2, -4, \ldots$, thus for $\Re(s) < 0$.

There is a pole at $s = 1$.

A.1.1 The Origins of the Functional Equation for Dirichlet's η Function

Euler in his "Remarques sur un beau rapport entre les series des puissances tant directes que reciproches"[1] writes the following functional equations

$$\frac{1-2^{n-1}+3^{n-1}-4^{n-1}+5^{n-1}-6^{n-1}+\cdots}{1-2^{-n}+3^{-n}-4^{-n}+5^{-n}-6^{-n}+\cdots} = -\frac{1\cdot2\cdot3\ldots.(n-1)(2^n-1)}{(2^{n-1}-1)\pi^n}\cos\left(\frac{n\pi}{2}\right)$$

$$\frac{1-3^{n-1}+5^{n-1}-7^{n-1}+\cdots}{1-3^{-n}+5^{-n}-7^{-n}+\cdots} = \frac{1\cdot2\cdot3\ldots.(n-1)(2^n)}{\pi^n}\sin\left(\frac{n\pi}{2}\right).$$

Then he finishes his work by proving that the above statements hold true for positive and negative whole numbers as well as for fractional values of n.

Nowadays we write with $s \in \mathbb{C}$:

$$\eta(1-s) = -\frac{(2^s-1)}{\pi^s(2^{s-1}-1)}\cos\left(\frac{\pi s}{2}\right)\Gamma(s)\eta(s) \tag{A.6}$$

which is the functional equation of Dirichlet's η function.

Hardy gave a proof for the case when s is replaced by $s + 1$ in the last equation:

$$\eta(-s) = 2\frac{\left(1-2^{-s-1}\right)}{1-2^{-s}}\pi^{-s-1}s\sin\left(\frac{\pi s}{2}\right)\Gamma(s)\eta(s+1) \tag{A.7}$$

From the relation $\eta(s) = \left(1-2^{1-s}\right)\zeta(s)$ one can show that η has zeroes at the points $s_k = 1 + \frac{2\pi i k}{\ln 2}$ for all $k \in \mathbb{Z} \setminus \{0\}$, e.g., $s_1 = 1 + 9.0647i$. For $k = 0$ one finds instead $\eta(1) = \ln 1 = 0.69315$. Remember that $\zeta(1) = \infty$.

When we write

$$\zeta(s) = \frac{\eta(s)}{1-2^{1-s}}$$

we realize that $\eta(s)$ as well as $\left(1-2^{1-s}\right)$ have the same zeroes s_k with $k = 1, 2, 3, \ldots \eta(s)$ is also zero at the points where $\zeta(s)$ is zero. These are the trivial zeroes $s = -2, -4, -6, \ldots$ such that

$$\eta(-2) = \eta(-4) = \eta(-6) = \cdots = 0.$$

Finally, η, like ζ, possesses the non-trivial zeroes within the critical strip $\{s \in \mathbb{C} | 0 < \Re(s) < 1\}$. The celebrated unproven Riemann hypothesis claims that all non-trivial zeros of ζ are located on the axis $\Re(s) = \frac{1}{2}$.

---

[1] Remarks on a beautiful relation between direct as well as reciprocal power series.

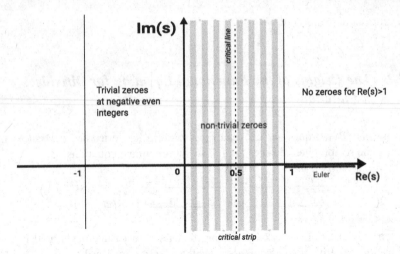

Fig. A.2 The behaviour of Riemann's ζ-function for real arguments

$\zeta(s)$ is a meromorphic function. Later we will meet Riemann's ξ function, $\xi(s) = \frac{1}{2}s(s-1)\pi^{-\frac{s}{2}}\Gamma\left(\frac{s}{2}\right)\zeta(s) \cdot \xi(s)$ is an entire function, it has non-trivial zeroes, however no trivial zeroes and no poles. Also: $\xi(s) = \xi(1-s)$ (Fig. A.2).

Table A.1 indicate that the Γ function and trigonometric factors in the functional equation ((A.1), (A.2), resp.) are tied to the trivial zeros and poles of the ζ function, but have no direct bearing on the distribution of the non-trivial zeroes, which is the most important feature of the ζ function for the purposes of analytic number theory, beyond the fact that they are symmetric about the real axis and the critical line $x = \frac{1}{2}$. Exponential functions such as 2^{s-1} or π^{-s} have neither zeroes nor poles. In particular the Riemann hypothesis is not going to be resolved just from further analysis of the Γ function.

Remarkable historical fact: Euler, in 1749 (110 years before Riemann!) discovered that the following series is convergent:

$$\phi(s) = \sum_{n=1}^{\infty} \frac{(-1)^{n+1}}{n^s} \tag{A.8}$$

This is also referred to as Dirichlet's η function. It is related to ζ by

$$\phi(s) = (1 - 2^{1-s})\zeta(s) \tag{A.9}$$

Within the critical strip $0 < s < 1$ we have:

$$\zeta(s) = \frac{2^{s-1}}{2^{s-1}-1}\phi(s) = \frac{1}{1-2^{1-s}}\phi(s)$$

Table A.1 Properties and special values of the Riemann ζ function

(a) A few values of $\zeta(s)$

s	$\zeta(s)$
$-2\mathbb{N}$	0
$-\mathbb{N}$	$\frac{-B_{n+1}}{n+1}$
-7	$\frac{1}{240}$
-5	$\frac{1}{252}$
-3	$\frac{1}{120}$
-1	$\frac{-1}{12}$
0	$-\frac{1}{2}$
$\frac{1}{2}$	-1.46035450
1	∞
$\frac{3}{2}$	2.6123753486
2	$\frac{\pi^2}{6} \approx 1.6449340$ (Euler, Basel)
$\frac{5}{2}$	1.3414872572
3	1.2020569
$\frac{7}{2}$	1.1267338673
4	$\frac{\pi^4}{90} \approx 1.082323233$

(b) Function properties

Function	Non-trivial zeroes	Trivial zeroes	Poles
$\zeta(s)$	Yes	$-2, -4, -6, \ldots$	1
$\zeta(1-s)$	Yes	$3, 5, \ldots$	0
$\sin\frac{\pi s}{2}$	No	$2\mathbb{N}$	No
$\cos\frac{\pi s}{2}$	No	$2\mathbb{N}+1$	No
$\sin\pi s$	No	$\mathbb{N}$	No
$\Gamma(S)$	No	No	$0, -1, -2, \ldots$
$\Gamma\left(\frac{S}{2}\right)$	No	No	$0, -2, -4, \ldots$
$\Gamma(1-S)$	No	No	$1, 2, 3, \ldots$
$\Gamma\left(\frac{1-s}{2}\right)$	No	No	$1, 3, 5, \ldots$
$\xi(S)$	Yes	No	No

$$= \frac{1}{1-2^{1-s}} \sum_{n=1}^{\infty} \frac{(-1)^{n+1}}{n^s}, \quad \Re(s) > 0,\ 1-2^{1-s} \neq 0. \tag{A.10}$$

From Euler we have

$$\frac{\phi(1-n)}{\phi(n)} = \frac{-(n-1)!(2^n-1)}{(2^{n-1}-1)\pi^n} \cos\left(\frac{n\pi}{2}\right), \tag{A.11}$$

and he furthermore says: "I shall hazard the following conjecture:

$$\frac{\phi(1-s)}{\phi(s)} = -\frac{\Gamma(s)(2^s - 1)\cos(\frac{\pi s}{2})}{(2^{s-1} - 1)\pi^s} \tag{A.12}$$

is true for all s". We know that $(\eta(s) =)\phi(s) = (1 - 2^{1-s})\zeta(s)$, which leads at once from (A.12) to

$$\zeta(1-s) = \frac{2}{(2\pi)^s}\Gamma(s)\zeta(s)\cos\left(\frac{\pi s}{2}\right), \quad \forall s \in \mathbb{C} \backslash 1 \tag{A.13}$$

and this is the famous functional equation which was proven by Riemann in 1859 (but it was conjectured by Euler in 1749!). It is probably correct to assume that Riemann was very familiar with Euler's contribution.

With the alternating Dirichlet series at hand we can already make an important statement regarding the zeroes of the ζ function within the critical strip $0 < \Re(s) = \sigma < 1$, which is important for the Riemann hypothesis, which claims that all non-trivial zeroes of ζ lie on the line with $\Re(s) = \frac{1}{2}$.

To show this we start with

$$\zeta(s) = \sum_{n=1}^{\infty} \frac{1}{n^s}, \quad s := \alpha + it \tag{A.14}$$

which is convergent for $\Re(s) > 1$, is a meromorphic function and has a pole at $s = 1$. Next let

$$n^s = n^{\sigma+it} = n^\sigma n^{it} = n^\sigma e^{it \ln n} = |n|^\sigma (\cos(t \ln n) + i \sin(t \ln n)) \tag{A.15}$$

from which immediately follows

$$\zeta(s) = \Re(\zeta(s)) + i\Im(\zeta(s)) = \sum_{n=1}^{\infty} \frac{1}{n^\sigma}[\cos(t \ln n) - i \sin(t \ln n)] \tag{A.16}$$

$$\Rightarrow \Re(\zeta(s)) = \sum_{n=1}^{\infty} n^{-\sigma} \cos(t \ln n) \tag{A.17}$$

$$\Im(\zeta(s)) = -\sum_{n=1}^{\infty} n^{-\sigma} \sin(t \ln n) \tag{A.18}$$

which are convergent for $\sigma > 1$, $t \in \mathbb{R}$. Next consider the Euler's ϕ function as given in (A.8), which is also known as Dirichlet's η function. An extension of the domain of ζ into the region of $0 < \sigma < 1$, i.e., into the critical strip, is obtained by rewriting (A.9) as

$$\zeta(s) = \frac{1}{1 - 2^{1-s}} \eta(s). \tag{A.19}$$

Note that only the critical strip is of importance for the Riemann hypothesis. Note further that η is convergent for $\sigma = \Re(s) > 0$ and that the following alternating harmonic series,

$$\eta(1) = 1 - \frac{1}{2} + \frac{1}{3} - \frac{1}{4} + \cdots = \ln 2 \approx 0.69315, \tag{A.20}$$

is obtained from

$$\ln(x + 1) = x - \frac{1}{2}x^2 + \frac{1}{3}x^3 - \cdots \quad -1 < x \le 1, \tag{A.21}$$

where x is assumed to be real. One may rewrite Dirichlet's η function in the following way:

$$\eta(s) = \sum_{n=1}^{\infty} \left(\frac{1}{(2n-1)^s} - \frac{1}{(2n)^s} \right). \tag{A.22}$$

From which one then obtains in a simple way (c.f. (A.17), (A.18)):

$$\Re(\eta(s)) = \sum_{n=1}^{\infty} [(2n-1)^{-\sigma} \cos(t \ln(2n-1)) - (2n)^{-\sigma} \cos(t \ln(2n))] \tag{A.23}$$

$$\Im(\eta(s)) = \sum_{n=1}^{\infty} [(2n)^{-\sigma} \sin(t \ln(2n)) - (2n-1)^{-\sigma} \sin(t \ln(2n-1))]. \tag{A.24}$$

Using $\cos x - \sin x = \sqrt{2}\sin\left(x + \frac{3}{4}\pi\right)$ one then obtains

$$\Re(\eta(s)) + \Im(\eta(s)) = \sqrt{2}\sum_{n=1}^{\infty} \left[(2n-1)^{-\sigma} \sin\left(t \ln(2n-1) + \frac{3}{4}\pi\right) \right.$$
$$\left. -(2n)^{-\sigma} \sin\left(t \ln(2n) + \frac{3}{4}\pi\right) \right] \ne 0 \; \forall \sigma \in (0, \frac{1}{2}), \forall t, \tag{A.25}$$

i.e., η possesses no roots on the left half of the critical strip, and because of the reflection formula (A.2) this holds true for the right half as well, i.e., they can only be on the critical line $\sigma = \frac{1}{2}$, which is the **Riemann hypothsis** (Fig. A.3).

Theorem *If* $\Re(s) = \sigma > 0$ *we have*

$$(1 - 2^{1-s})\zeta(s) = \eta(s) = \sum_{n=1}^{\infty} \frac{(-1)^{n-1}}{n^s}, \tag{A.26}$$

Fig. A.3 A closer look at the behavior of ζ. Referring to A.3b we have $\left|\zeta\left(\frac{1}{2}-\sigma\right)\right| > \left|\zeta\frac{1}{2}+\sigma\right|$ or $\left|\zeta\left(\frac{1}{2}-\sigma\right)\right| > \left|\zeta\left(\frac{1}{2}\right)\right|$. No zeroes of ζ on the left half and right half of the critical strip, which is equivalent to Riemann's hypothesis

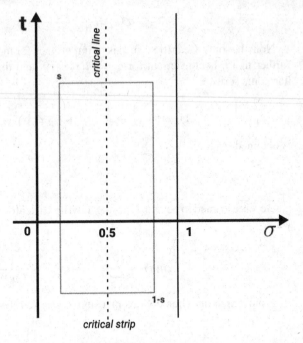

(a) The argument

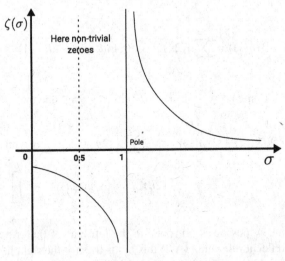

(b) The Function

which implies that $\zeta(s) < 0$ if s is real and $0 < s < 1$.

Proof. First assume that $\sigma > 1$ (Euler: $\Re(s) > 1$). Then we have

$$(1 - 2^{1-s})\zeta(s) = \sum_{n=1}^{\infty} \frac{1}{n^s} - 2 \sum_{n=1}^{\infty} \frac{1}{(2n)^s}$$

$$= (1 + 2^{-s} + 3^{-s} + \cdots) - 2(2^{-s} + 4^{-s} + 6^{-s} + \cdots)$$

$$= 1 - 2^{-s} + 3^{-s} - 4^{-s} + \cdots = \text{alternating } \zeta \text{ function},$$

which proves (A.26) for $\Re(s) = \sigma > 1$. However, if $\sigma > 0$ the series on the right converges, thus (A.26) also holds for $\sigma > 0$ by analytic continuation, i.e., when s is real then the sum in (A.26) is an alternating series with a positive limit.

If $0 < s < 1$, then the factor $1 - 2^{1-s}$ becomes negative. Hence $\zeta(s)$ is also negative (has no zeroes!) in $0 < s < 1$.   □

Note that $\eta(1) = \cdots = \ln 2 \approx 0.69315$ (c.f. (A.20)) while $\zeta(1) = \infty$, that is, $s = 1$ is a pole of the meromorphic function ζ. Furthermore we have

$$\zeta(0) = -\frac{1}{2}. \tag{A.27}$$

Proof. Starting with the functional equation

$$\Gamma\left(\frac{s}{2}\right)\pi^{-\frac{s}{2}}\zeta(s) = \Gamma\left(\frac{1-s}{2}\right)\pi^{-\frac{1-s}{2}}\zeta(1-s) \tag{A.28}$$

solve for $\zeta(s)$ to obtain

$$\zeta(s) = \pi^{\frac{s}{2}}\pi^{-\frac{1-s}{2}}\Gamma\left(\frac{1-s}{2}\right)\frac{\zeta(1-s)}{\Gamma\left(\frac{s}{2}\right)}$$

$$s \to 0 : \zeta(0) = \pi^{\frac{-1}{2}}\Gamma\left(\frac{1}{2}\right)\lim_{s\to 0}\frac{\zeta(1-s)}{\Gamma\left(\frac{s}{2}\right)}.$$

Since the residues of ζ at $s = 1$ and of Γ at $s = 0$ are both 1, i.e.,

$$\zeta(s) = \frac{1}{s-1} + \cdots, \quad \Gamma(s) = \frac{1}{s} + \cdots, \tag{A.29}$$

we have

$$\zeta(1-s) = -\frac{1}{s} + \cdots, \quad \Gamma\left(\frac{s}{2}\right) = \frac{2}{s} + \cdots \tag{A.30}$$

and therefore

$$\lim_{s \to 0} \frac{\zeta(1-s)}{\Gamma\left(\frac{s}{2}\right)} = \lim_{s \to 0} -\frac{\frac{1}{s} + \cdots}{\frac{2}{s} + \cdots} = -\frac{1}{2} \tag{A.31}$$

from which follows, using $\Gamma\left(\frac{1}{2}\right)$

$$\zeta(0) = \pi^{-\frac{1}{2}}\pi^{\frac{1}{2}}\left(-\frac{1}{2}\right) = -\frac{1}{2} \Rightarrow \zeta(0) = -\frac{1}{2}. \tag{A.32}$$

$\square$

From the Eqs. (4.31), (4.32) we have

$$t, x, \psi(x), \ln(x) \in \mathbb{R}.$$

Therefore $\Im\xi\left(\frac{1}{2} + it\right) = 0$, i.e., $\xi\left(\frac{1}{2} + it\right) \equiv \Xi(t) \in \mathbb{R}$ and thus

$$\Xi(t) = \xi\left(\frac{1}{2} + it\right) = -\frac{t^2 + \frac{1}{4}}{2(\sqrt{\pi})^{\frac{1}{2} + it}}\Gamma\left(\frac{1}{4} + \frac{it}{2}\right)\zeta\left(\frac{1}{2} + it\right)$$

$$\xi\left(\frac{1}{2}\right) = -\frac{1}{8\pi^{\frac{1}{4}}}\Gamma\left(\frac{1}{4}\right)\zeta\left(\frac{1}{2}\right) \approx 0.4971207781 =: a_0$$

$$\zeta\left(\frac{1}{2}\right) \approx -1.4603545088$$

$$\Gamma\left(\frac{1}{4}\right) = \sqrt{2\bar{\omega}2\pi} \approx 3.6256099082$$

where in the last equation $\bar{\omega}$ is the so-called Gaussian lemniscate constant.

Some special values:

$$\xi(0) = \xi(1) = -\zeta(0) = \frac{1}{2}. \tag{A.33}$$

Proof. Using $\xi(s) = \frac{1}{2}s(s-1)\pi^{-\frac{s}{2}}\Gamma\left(\frac{s}{2}\right)\zeta(s)$ as well as $\Gamma\left(1 + \frac{s}{2}\right) = \frac{s}{2}\Gamma\left(\frac{s}{2}\right)$ we obtain

$$\xi(s)|_{s=0} = (s-1)\pi^{\frac{-s}{2}}\Gamma\left(1 + \frac{s}{2}\right)\zeta(s)|_{s=0} \Leftrightarrow \xi(s) = -1 \cdot 1 \cdot \Gamma(1) \cdot \zeta(0) = \frac{1}{2} \tag{A.34}$$

Thus

$$\xi(0) = \frac{1}{2}.$$

In a similar manner, utilizing the reflection property $\xi(s) = \xi(1-s)$:

$$\xi(s) = (-s)\pi^{-\frac{1}{2}(1-s)}\Gamma\left(\frac{3}{2} - \frac{s}{2}\right)\zeta(1 - s)$$

$$\Rightarrow \xi(1) = -1 \cdot 1 \cdot \Gamma(1) \cdot \zeta(0) = \frac{1}{2}$$

$$\Rightarrow \xi(1) = \frac{1}{2} \tag{A.35}$$

□

A.1.2 Riemann's Functional Equation

$$\pi^{-\frac{s}{2}}\Gamma\left(\frac{s}{2}\right)\zeta(s) = \pi^{-\frac{1-s}{2}}\Gamma\left(\frac{1-s}{2}\right)\zeta(1 - s) \tag{A.36}$$

whose symmetry is obvious when $s \to 1 - s$ is substituted into both sides of the equation.

Proof. Starting with Euler's Γ function

$$\Gamma(s) = \int_0^\infty t^{s-1}e^{-t}dt. \tag{A.37}$$

Using $s \to \frac{s}{2}$, the above results in

$$\Gamma\left(\frac{s}{2}\right) = \int_0^\infty t^{\frac{s}{2}-1}e^{-t}dt. \tag{A.38}$$

Next one can use the substitution $t = \pi n^2 x (dt = \pi n^2 dx)$ to obtain

$$\Gamma\left(\frac{s}{2}\right) = \int_0^\infty (\pi n^2 x)^{\frac{s}{2}-1}e^{-\pi n^2 x}\pi n^2 dx$$
$$\pi^{-\frac{s}{2}}\Gamma\left(\frac{s}{2}\right)\frac{1}{n^s} = \int_0^\infty x^{\frac{s}{2}-1}e^{-\pi n^2 x}dx$$

Summation over n yields

$$\sum_{n=1}^\infty \pi^{-\frac{s}{2}}\Gamma\left(\frac{s}{2}\right)\frac{1}{n^s} = \sum_{n=1}^\infty \int_0^\infty x^{\frac{s}{2}-1}e^{-\pi n^2 x}dx$$

$$\pi^{-\frac{s}{2}}\Gamma\left(\frac{s}{2}\right)\sum_{n=1}^\infty \frac{1}{n^s} = \int_0^\infty x^{\frac{s}{2}-1}\sum_{n=1}^\infty e^{-\pi n^2 x}dx$$

$$\pi^{-\frac{s}{2}}\Gamma\left(\frac{s}{2}\right)\zeta(s) = \int_0^\infty x^{\frac{s}{2}-1}\underbrace{\sum_{n=1}^\infty e^{-\pi n^2 x}}_{\text{closely related to Jacobi } \vartheta \text{ func.}} dx$$

$$\vartheta(x) = \sum_{n\in\mathbb{Z}} e^{-\pi n^2 x} = 1 + 2\sum_{n=1}^{\infty} e^{-\pi n^2 x} = 1 + 2\psi(x), x > 0.$$

$$\Rightarrow \int_0^\infty x^{\frac{s}{2}-1} \sum_{n=1}^{\infty} e^{-\pi n^2 x} dx = \int_0^\infty x^{\frac{s}{2}-1} \psi(x) dx.$$

Split the integral of the r.h.s into two parts:

$$\int_0^\infty x^{\frac{s}{2}-1}\psi(x)dx = \int_1^\infty x^{\frac{s}{2}-1}\psi(x)dx + \int_0^1 x^{\frac{s}{2}-1}\psi(x)dx. \tag{A.39}$$

Look at $\vartheta(x) = \frac{1}{\sqrt{x}}\vartheta\left(\frac{1}{x}\right)$ or $2\psi(x) + 1 = \frac{1}{\sqrt{x}}(1 + \psi(\frac{1}{x}))$. The Eqs. (4.5)ff. in the body of the paper are

$$\psi(x) = \frac{1}{\sqrt{x}}\psi\left(\frac{1}{x}\right) - \frac{1}{2} + \frac{1}{2\sqrt{x}}$$

$$\int_0^1 x^{\frac{s}{2}-1}\psi(x)dx = \int_0^1 x^{\frac{1}{2}-1}\left(\frac{1}{\sqrt{x}}\psi\left(\frac{1}{x}\right) + \frac{1}{2\sqrt{x}} - \frac{1}{2}\right)dx$$

$$= \int_0^1 \left(x^{\frac{s}{2}-\frac{3}{2}}\psi\left(\frac{1}{x}\right) + \frac{1}{2}\left(x^{\frac{s}{2}-\frac{3}{2}} - x^{\frac{s}{2}-1}\right)\right)dx$$

$$= \int_0^1 x^{\frac{s-3}{2}}\psi\left(\frac{1}{x}\right)dx + \frac{1}{2}\left[\frac{1}{\frac{s}{2}-\frac{1}{2}}x^{\frac{s}{2}-\frac{1}{2}} - \frac{1}{\frac{s}{2}}x^{\frac{s}{2}}\right]_0^1$$

$$= \int_0^1 x^{\frac{s}{2}-\frac{3}{2}}\psi\left(\frac{1}{x}\right)dx + \frac{1}{s(s-1)}$$

$$\overset{(*)}{=} \int_\infty^1 \left(\frac{1}{y}\right)^{\frac{s}{2}-\frac{3}{2}}\psi(y)\left(-\frac{1}{y^2}\right)dy + \frac{1}{s(s-1)}$$

$$\overset{y\to x}{=} \int_1^\infty \left(\frac{1}{x}\right)^{\frac{s}{2}-\frac{3}{2}}\psi(x)\frac{dx}{x^2} + \frac{1}{s(s-1)}$$

$$\Rightarrow \int_0^1 x^{\frac{s}{2}-1}\psi(x)dx = \int_1^\infty x^{-\frac{s}{2}-\frac{1}{2}}\psi(x)dx + \frac{1}{s(s-1)}$$

$$\int_0^\infty x^{\frac{s}{2}-1}\psi(x)dx = \int_1^\infty x^{\frac{s}{2}-1}\psi(x)dx + \int_0^1 x^{\frac{s}{2}-1}\psi(x)dx$$

$$= \int_1^\infty x^{\frac{s}{2}-1}\psi(x)dx + \int_1^\infty x^{\frac{-s}{2}-\frac{1}{2}}\psi(x)dx + \frac{1}{s(s-1)}$$

$$= \int_1^\infty \left(x^{\frac{s}{2}-1} + x^{-\frac{s}{2}-\frac{1}{2}}\right)\psi(x)dx + \frac{1}{s(s-1)},$$

where in (*) the substitution $x = \frac{1}{y}$, $dx = -\frac{1}{y^2}dy$, $\int_0^1 \to \int_\infty^1$ was used. Recall that we started with $\pi^{-\frac{s}{2}}\Gamma\left(\frac{s}{2}\right)\zeta(s) = \int_0^\infty x^{\frac{s}{2}-1}\psi(x)dx$ and arrived at

$$\pi^{-\frac{s}{2}}\Gamma\left(\frac{s}{2}\right)\zeta(s) = \int_1^\infty \left(x^{\frac{s}{2}} + x^{\frac{1-s}{2}}\right)\frac{\psi(x)}{x}dx - \frac{1}{s(s-1)}. \tag{A.40}$$

Note that the last term carries the pole of Γ at $s = 0$ and of ζ at $s = 1$. Note further that the r.h.s. does not change under $s \to 1 - s$, which implies Riemann's functional equation

$$\pi^{-\frac{s}{2}}\Gamma\left(\frac{s}{2}\right)\zeta(s) = \pi^{-\frac{1-s}{2}}\Gamma\left(\frac{1-s}{2}\right)\zeta(1-s).$$

Riemann used 4–5 lines to derive this relation!   □

In (A.40) we used

$$x^{\frac{s}{2}} = x^{\frac{\sigma+it}{2}} = e^{\frac{\sigma \ln(x)}{2} + i\frac{t}{2}\ln(x)} = e^{\frac{\sigma \ln(x)}{2}}\left[\cos\left(\frac{t}{2}\ln(x)\right) + i\sin\left(\frac{t}{2}\ln(x)\right)\right]$$

$$x^{\frac{1-x}{2}} = e^{\frac{(1-\sigma)\ln(x)}{2}}\left[\cos\left(\frac{t}{2}\ln(x)\right) - i\sin\left(\frac{t}{2}\ln(x)\right)\right]$$

$$x^{\frac{s}{2}} = x^{\frac{1}{2}(1-s)} = \left(e^{\frac{\sigma \ln(x)}{2}} + e^{\frac{(1-\sigma)\ln(x)}{2}}\right)\cos\left(\frac{t}{2}\ln(x)\right)$$

$$\underset{=}{y=\frac{t}{2}\ln(x)}\left(e^{\sigma\frac{y}{t}} + e^{(1-\sigma)\frac{y}{t}}\right)\cos(y)$$

$$\underset{=}{R.H.::\sigma=\frac{1}{2}}\left(e^{\frac{y}{2t}} + e^{\frac{y}{2t}}\right)\cos(y) = 2e^{\frac{y}{2t}}\cos(y)$$

$$= 2e^{\frac{1}{4}\ln(x)}\cos(y) = 2x^{\frac{1}{4}}\cos\left(\frac{t}{2}\ln(x)\right)$$

and whose imaginary part vanishes for $\sigma = \frac{1}{2}$. Thus

$$\Xi(t) := \xi\left(\frac{1}{2} + it\right) = \frac{1}{2} + \frac{1}{2}s(s-1)\int_1^\infty \psi(x) \cdot 2 \cdot e^{\frac{1}{4}\ln(x)}\cos\left(\frac{t}{2}\ln(x)\right)\frac{dx}{x} \tag{A.41}$$

is a real function, which is mentioned in Riemann's Berlin paper on p. 147 as

$$\Xi(t) = \frac{1}{2} - \left(t^2 + \frac{1}{4}\right)\int_1^\infty \psi(x)x^{-\frac{3}{4}}\cos\left(\frac{t}{2}\ln(x)\right)dx; \tag{A.42}$$

furthermore,

$$\Im\xi\left(\frac{1}{2} + it\right) = 0, \Rightarrow \xi\left(\frac{1}{2} + it\right) = \Xi(t) \in \mathbb{R}. \tag{A.43}$$

A.1.3   What Is a Function?

Why is $1+2+3+4+\cdots = -\frac{1}{12}$. a regularized value? A normal reaction to this result:

This is not a true result. It is hogwash to say that $1+2+3+\cdots$ has a finite value, as long as one does not specify what a function is (the concept of a function) and how it is calculated, i.e., which representation is chosen, what its domain of definition is, etc.

The following two statements are, however, true:

$$1+2+3+4+\cdots \to \infty, \text{ i.e., divergent}$$

$$\zeta_{\text{Riemann}}(-1) = -\frac{1}{12}.$$

Question: In which representation is the latter statement true? We need a more general understanding of a function as well as the representation in which the value of the function is calculated.

It is well known that a function can have several different representations, e.g., taking the sine function:

$$f(z) = \begin{cases} \sin(z) & \\ \frac{e^{iz}-e^{iz}}{2i} & \text{Euler} \\ z - \frac{z^3}{3!} + \frac{z^5}{5!} - \cdots & \text{Taylor expansion} \\ z \prod_{n=1}^{\infty} \left(1 - \frac{z^2}{\pi^2 n^2}\right) & \text{Product expansion} \end{cases} \tag{A.44}$$

The Taylor expansion is an infinite-sum expansion of the sine function, one needs only powers of z. The product expansion of the sine function needs all the infinitely many zeroes of the sine function. One sees that there are <u>many</u> different ways to write a single function (e.g., sine), i.e., many different expressions for performing various calculations!

What does all of this mean for the zeta function? Let's start with Euler's definition (1737):

$$\zeta(s) = \sum_{n=1}^{\infty} \frac{1}{n^s}, \quad s > 1$$

$$= 1 + \frac{1}{2^s} + \frac{1}{3^s} + \cdots, \quad s > 1 \text{ for convergence,}$$

which is a sum of reciprocal powers of integers. Evidently substituting <u>negative</u> numbers for s is not allowed, not even $s = 1$ is permitted.

If one ignores the convergence condition $s > 1$, then one can write

$$\zeta_{\text{Euler}}(-1) = 1 + \frac{1}{2^{-1}} + \frac{1}{3^{-1}} + \cdots = 1 + 2 + 3 + 4 + \cdots, \tag{A.45}$$

which is pure nonsense, because it is not correctly defined. $s = -1$ is simply not allowed in Euler's definition (representation) of the zeta function, which is only defined on the real axis $1 < x \equiv s$. But there is another representation attributed to Riemann, which can be extended into the whole complex plane, $s \in \mathbb{C} \setminus \{0, 1\}$, i.e., including the value $\Re(s) = -1$.

$$\zeta(s) = \begin{cases} \zeta_E(s) = \sum_{n=1}^{\infty} \frac{1}{n^s} & \Re(s) > 1, \text{Euler}(1797) \\ \zeta_R(s) = 2^s \pi^{s-1} \sin\left(\frac{\pi s}{2}\right)\Gamma(1-s)\zeta(1-s) & s \in \mathbb{C}\setminus\{0, 1\}, \text{Riemann (1859)} \end{cases} \tag{A.46}$$

Note that the latter function is not given as a series but as a meromorphic function. In Riemann's representation we obtain

$$\zeta_R(-1) = 2^{-1}\pi^{-2}\sin\left(\frac{-\pi}{2}\right)\Gamma(1-(-1))\zeta(1-(-1))$$

$$= 2^{-1}\pi^{-2}(-1)\Gamma(2)\zeta(2)$$

$$= 2^{-1}\pi^{-2}(-1) \cdot 1 \cdot \frac{\pi^2}{6} = -\frac{1}{12},$$

where in the third equality we used $\Gamma(2) = (2-1)! \cdot 1 = 1$, $\zeta(s) = 1 + \frac{1}{2^2} + \frac{1}{3^2} + \cdots = \frac{\pi^2}{6}$.

This is a true statement in Riemann's zeta-function representation

$$-\frac{1}{12} = \zeta_R(-1) \neq \zeta_E(-1) = \sum_{n=1}^{\infty} \frac{1}{n^s}|_{s=-1} \stackrel{*}{=} 1 + 2 + 3 + 4 + \cdots \tag{A.47}$$

where as Euler's representation is not defined for $s = -1$.

The prime number counting function $\pi(x)$.

Claim:

$$\frac{\ln \zeta(s)}{s} = \int_2^{\infty} \frac{\pi(x)}{x(x^s - 1)}dx, \quad s > 1$$

$$\zeta(s) = \prod_{p \in \text{primes}} \frac{1}{1 - p^{-s}}, \quad s > 1$$

$$\ln \zeta(s) = \ln \prod_{p \in \text{primes}} \frac{1}{1 - p^{-s}} = \sum_{p \in \text{primes}} \ln \frac{1}{1 - p^{-s}} \tag{A.48}$$

where $\pi(x)$ is the number of primes smaller than x. Replacing the summation over the primes by a summation over all integers yields

$$\ln \zeta(s) = \sum_{n=2}^{\infty} \{\pi(n) - \pi(n-1)\} \ln \frac{1}{1 - n^{-s}} \qquad (A.49)$$

where

$$\pi(n) - \pi(n-1) = \begin{cases} 1, & n \in \text{primes} \\ 0, & \text{else} \end{cases}$$

projects out the primes, e.g.,

$$\pi(2) - \pi(1) = 1 - 0 = 1$$
$$\pi(3) - \pi(2) = 2 - 1 = 1$$
$$\pi(4) - \pi(3) = 2 - 2 = 0$$
$$\vdots$$

$$(A.49) \Rightarrow \ln \zeta(s) = \sum_{n=2}^{\infty} \pi(n) \ln \frac{1}{1 - n^{-s}} - \sum_{n=2}^{\infty} \pi(n-1) \ln \frac{1}{1 - n^{-s}}$$

$$= \sum_{n=2}^{\infty} \pi(n) \ln \frac{1}{1 - n^{-s}} - \sum_{n=2}^{\infty} \pi(n) \ln \frac{1}{1 - (n+1)^{-s}}$$

$$= \sum_{n=2}^{\infty} \pi(n)(\ln(1 - (n+1)^{-s}) - \ln(1 - n^{-s})). \qquad (A.50)$$

Now use

$$\frac{d}{dx} \ln(1 - x^{-s}) = \frac{1}{1 - x^{-s}}(sx^{-s-1}) = \frac{s}{x(x^s - 1)}. \qquad (A.51)$$

Integrate both sides to obtain

$$\ln(1 - x^{-s}) = s \int \frac{1}{x(x^s - 1)} dx + C \qquad (A.52)$$

and use it in (A.50), whilst converting the indefinite integral into one over $[n, n+1]$:

$$\ln \zeta(s) = \sum_{n=2}^{\infty} \underbrace{\pi(n)}_{\text{const.under integral}} \int_{n}^{n+1} \frac{s}{x(x^s - 1)} dx$$

$$= \sum_{n=2}^{\infty} \int_{n}^{n+1} \frac{s\pi(x)}{x(x^s - 1)} dx \quad n : 2 \to 3, 3 \to 4, \ldots$$

$$\ln \zeta(s) = \int_{2}^{\infty} \frac{s\pi(x)}{x(x^s - 1)} dx$$

or

$$\frac{\ln \zeta(s)}{s} = \int_2^\infty \frac{\pi(x)}{x(x^s - 1)} dx.$$

This concludes the proof.

For s > 1 there are no non-trivial zeroes of ζ . Such are located in the critical strip $0 < \Re(s) = \sigma < 1$. The Riemann Hypothesis states that $\sigma = \frac{1}{2}$ for all zeroes of the ζ function. Hence the formula (A.48) is not applicable and we have to make an analytic continuation into the entire complex s plane.

A.2 Appendix

A.2.1 Remarks on the Oscillatory Behavior of the Prime Counting Function

Beginning with Gauss' observation that $Li(x)$ sets an upper bound for the prime number staircase $\pi(x)$, at least up to $x = 10^8$, it was Bernhard Riemann who suggested that instead of $Li(x)$ it should be the weighted sum $\sum_{n=1}^\infty \pi(x^{\frac{1}{n}})$ and not $\pi(x)$ alone. This means that the prime counting number $\Pi(x)$ should be replaced by $Li(x)$:

$$\Pi(x) = \sum_{n=1}^\infty \frac{1}{n} \pi\left(x^{\frac{1}{n}}\right) \rightarrow L_i(x) \approx \pi(x) + \frac{1}{2}\pi\left(x^{\frac{1}{2}}\right) + \frac{1}{3}\pi\left(x^{\frac{1}{3}}\right) + \cdots$$

Application of the Möbius inversion formula then leads to

$$\pi(x) \approx L_i(x) - \frac{1}{2}L_i\left(x^{\frac{1}{2}}\right) - \frac{1}{3}L_i\left(x^{\frac{1}{3}}\right) \cdots$$

Here we meet Gauss' dominant term $L_i(x)$ augmented by an infinite series of refinements. Hence we have the following approximation for $\pi(x)$ $(\pi \rightarrow R, \Pi \rightarrow L_i)$:

$$R(x) = \sum_{n=1}^\infty \frac{\mu(n)}{n} L_i\left(x^{\frac{1}{n}}\right)$$

instead of

$$\pi(x) = \sum_{n=1}^\infty \frac{\mu(n)}{n} \Pi\left(x^{\frac{1}{n}}\right). \tag{A.53}$$

(R stands for the first letter in "Riemann.")

Now we can read off the Riemann-modified value for $\pi(x)$. Then, with ϱ denoting the non-trivial zeros of the ζ function, we need to calculate

$$\pi = R(x^1) - \sum_{\varrho} R(x^{\varrho}),$$

with

$$R(x^1) = \sum_{n=1}^{\infty} \frac{\mu(n)}{n} L_i\left(x^{\frac{1}{n}}\right) \text{ and } R(x^{\varrho}) = \sum_{n=1}^{\infty} \frac{\mu(n)}{n} L_i\left(x^{\frac{\varrho}{n}}\right) \qquad (A.54)$$

As further approximations, we limit ourselves to

$$R(x^1) = R(x) \approx L_i(x) - \frac{1}{2} L_i\left(x^{\frac{1}{2}}\right) \text{ with } L_i\left(x^{\frac{1}{2}}\right) \approx \frac{2\sqrt{x}}{\ln x},$$

and $R(x^{\varrho})$ should be approximated by the first term $n = 1$ in (A.54), i.e.,

$$R(x^{\varrho}) \approx L_i(x^{\varrho}) = L_i(e^{\varrho \ln x}) = E_i(\varrho \ln x). \qquad (A.55)$$

These formulae allow us to calculate the approximate value

$$\pi(x) \approx L_i(x) - \frac{1}{2} \frac{2\sqrt{x}}{\ln x} - \sum_{\varrho} \{L_i(x^{\varrho}) + L_i(x^{\tilde{\varrho}})\}, \qquad (A.56)$$

where we took care of the non-trivial ζ zeros and its mirror values.

With $z = \varrho \ln x$ and $\varrho = \frac{1}{2} + i\gamma$, we can write for (A.55)

$$E_i(z) \sim \frac{e^z}{z}\left(1 + \frac{1!}{z} + \frac{2!}{z^2} + \cdots\right),$$

which, when approximated by the first term, yields

$$E_i\left(\left(\frac{1}{2} + i\gamma\right)\ln x\right) \approx \frac{e^{\varrho \ln x}}{\varrho \ln x} = \frac{x^{\frac{1}{2}+i\gamma}}{\left(\frac{1}{2} + i\gamma\right)\ln x}$$

Hence, when adding the mirror terms in (A.56), we obtain

$$E_i\left(\left(\frac{1}{2} + i\gamma\right)\ln x\right) + E_i\left(\left(\frac{1}{2} - i\gamma\right)\ln x\right) \approx \frac{\sqrt{x}}{\ln x}\left(\frac{e^{i\gamma \ln x}}{\frac{1}{2} + i\gamma} + \frac{e^{-i\gamma \ln x}}{\frac{1}{2} - i\gamma}\right)$$

$$\approx \frac{\sqrt{x}}{\ln x} \frac{e^{i\gamma \ln x - \arg \varrho} + e^{-i\gamma \ln x + \arg \varrho}}{|\varrho|}$$

since $\varrho = \frac{1}{2} + i\gamma = |\varrho|e^{i\,\arg\varrho}$ $\bar{\varrho} = |\varrho|e^{-i\,\arg\varrho}$.

This yields Riesel's approximation (2.30) in his book, "Prime Numbers and Computer Methods for Factorization":

$$R(x^{\varrho}) + R(x^{\bar{\varrho}}) \approx \frac{\sqrt{x}}{\ln x} 2Re\left(e^{i\gamma lnx - i\,\arg\varrho}\right)$$

$$\approx \frac{2\sqrt{x}}{|\varrho|\ln x}\cos(Im\varrho\,\ln x - \arg\varrho)\quad Im\varrho = \gamma;$$

$$\arg\varrho = \frac{\pi}{2}(\text{Riemann}); |\varrho|^2 = \frac{1}{4} + \gamma^2; |\varrho| \to \gamma\; for\; \gamma \gg 1.$$

$$\approx 2\frac{\sqrt{x}}{\ln x}\frac{\sin(\gamma \ln x)}{\gamma}$$

Finally we end up with

$$\pi(x) \approx \int_2^x \frac{dt}{t} - \frac{\sqrt{x}}{\ln x}\left(1 + 2\sum_{k=1}^{\infty}\frac{\sin(\gamma_k \ln x)}{\gamma_k}\right).\gamma_k\; \text{Riemann's}\; \zeta\; \text{zeros}\quad (A.57)$$

$$Li(x) = \int_2^x \frac{dt}{t}.$$

This shows up in the paper, "Prime Races" by Granville and Martin in the form

$$\frac{\int_2^x \frac{dt}{t} - \pi(x)}{\frac{\sqrt{x}}{\ln x}} \approx \left(1 + 2\sum_{\gamma}\frac{\sin(\gamma \ln x)}{\gamma}\right).$$

The numerator on the left-hand side of this formula is the error term when comparing the Gauss prediction $L_i(x)$ with the actual count $\pi(x)$ for the number of primes up to x.

The right-hand side is the superposition of sine functions which resembles the Fourier transform of the prime number step function $\pi(x)$ with the numbers γ_k acting as "frequencies" inside the sine and the reciprocals as the "amplitude," where $k = 1, 2, 3, \ldots$ characterize the first, second, third, ... harmonic, numerated by the increasing positive nontrivial zeros of Riemann's ζ function.

Note: Nowhere do we talk about prime numbers in the oscillatory part of $\pi(x)$!

There exists another (easier) equivalent way to approximate the prime number step function. So far we counted the number of primes up to x by

$$\pi(x) = \sum_{p<x} 1,$$

i.e., where the step function increases by 1 if a new prime number is reached. It was the Russian mathematician Chebyshev who modified this kind of counting prime numbers by introducing the weighted prime number function

$$\psi(x) = \sum_{p<x} lnp,$$

i.e., $\psi(x)$ is raised by lnp every time a new prime number power is reached, e.g.,

$$\psi(10) = (ln2 + ln3 + ln5 + ln7) + (ln2) + (ln3)$$
$$= 2ln2 + 2ln3 + ln5 + ln7 = 7.1388$$
$$\psi(20) = (ln2 + ln3 + ln5 + ln7 + ln11 + ln13 + ln17 + ln19)$$
$$+ (ln2 + ln3) + (ln2) + (ln2)$$
$$= 4ln2 + 2ln3 + ln5 + ln7 + ln11 + ln13 + ln17 + ln19 = 19.2656$$
$$\psi(30) = 4ln2 + 3ln3 + 2ln5 + ln7 + \cdots + ln29 = 28.4765.$$

Here are a few more values for $\psi(x)$:

x	100	200	300	400	500	600	700	800	900	1000
$\psi(x)$	96.4	206.1	299.2	399.8	501.7	593.9	699.0	792.7	897.2	996.9

These numerical examples show convincingly that $\psi(x)$ lies very close to x. If we can prove this fact, we can conclude that the statement $\psi(x) \sim x$ is equivalent to the prime number theorem:

$$\frac{\pi(x)}{\frac{x}{lnx}} \text{ equivalent to } \frac{\psi(x)}{x} \text{ (for large } x).$$

Although the proof that follows is based on ideas by Chebyshev, it is provided by Mangoldt's famous formula (6.7):

$$\psi(x) = \sum_{p^n \leq x} lnp = \frac{1}{2\pi i} \int_{a-i\infty}^{a+i\infty} \left(-\frac{\zeta'(x)}{\zeta(x)} \frac{x^z}{z} \right) dz$$

which resulted in (6.9):

$$\psi(x) = x - \ln(2\pi) - \frac{1}{2} \ln\left(1 - \frac{1}{x^2}\right) - \sum_{\zeta(\varrho)=0} \frac{x^\varrho}{\varrho}$$

where x is essentially augmented by the last (oscillatory) term.

So let us compute the oscillatory part of $\psi(x)$:

$$\sum_{\zeta(\varrho)=0 \text{ mirror zeros}}' \frac{x^\varrho}{\varrho} = \sum \left(\frac{x^\varrho}{\varrho} + \frac{x^{\bar{\varrho}}}{\bar{\varrho}} \right), \frac{x^\varrho}{\varrho} + \frac{x^{\bar{\varrho}}}{\bar{\varrho}} = 2Re\frac{x^\varrho}{\varrho}, \varrho = \frac{1}{2} + i\gamma, \gamma > 0.$$

$$\frac{x^\varrho}{\varrho} + \frac{x^{\bar\varrho}}{\bar\varrho} = \frac{x^{1/2+i\gamma}}{\varrho} + \frac{x^{1/2-i\gamma}}{\bar\varrho} = x^{1/2}\left(\frac{e^{i\gamma lnx}}{\varrho} + \frac{e^{-i\gamma lnx}}{\bar\varrho}\right)$$

$$= \sqrt{x}\left(\frac{\bar\varrho e^{i\gamma lnx} + \varrho e^{-i\gamma lnx}}{|\varrho|^2}\right)$$

$$= \frac{\sqrt{x}}{|\varrho|^2}\left(|\varrho|e^{-i\,arg\,\varrho+i\gamma lnx} + |\varrho|e^{i\,arg\,\varrho-i\gamma lnx}\right)$$

$$= \frac{\sqrt{x}}{|\varrho|}\left(e^{i\gamma lnx-i\,arg\,\varrho} + e^{-i\gamma lnx+i\,arg\,\varrho}\right)$$

$$2Re\frac{x^\varrho}{\varrho} = 2\frac{\sqrt{x}}{|\varrho|}Re\left(e^{i(\gamma lnx-arg\,\varrho)}\right) = 2\frac{\sqrt{x}}{|\varrho|}\cos(\gamma lnx - arg\,\varrho)$$

$$\approx 2\sqrt{x}\frac{\sin(\gamma lnx)}{\gamma}.$$

Finally—with no reference to prime numbers!—we arrive at

$$\psi(x) = x - ln2\pi - \frac{1}{2}\ln\left(1 - \frac{1}{x^2}\right) - \sum_{k=1}^{\infty}\frac{2}{|\varrho_k|}\sqrt{x}\cos(Im\varrho_k lnx - arg\,\varrho_k)$$

(A.58)

or for

$$\gamma_k \gg 1 : \psi(x) = x - ln2\pi - \frac{1}{2}\ln\left(1 - \frac{1}{x^2}\right) - \sum_{k=1}^{\infty}\frac{2\sqrt{x}}{\gamma_k}\sin(\gamma_k lnx) \quad \text{(A.59)}$$

Comparing this with our former result for $\pi(x)$, we indeed see that $\frac{\pi(x)}{\frac{x}{lnx}}$ and $\frac{\psi(x)}{x}$ have the same asymptotic limit.

So instead of calculating Riemann's and von Mangoldt's (Chebyshev's) more exact expressions,

$$\pi(x) = \sum_{n=1}^{\infty}\frac{\mu(n)}{n}L_i\left(x^{\frac{1}{n}}\right) - \sum_{n=1}^{\infty}\frac{\mu(n)}{n}\sum_{\varrho}\left(L_i x^{\frac{\varrho}{n}}\right) - ln2\sum_{n=1}^{\infty}\frac{\mu(n)}{n}$$

$$+ \sum_{n=1}^{\infty}\frac{\mu(n)}{n}\int_{x^{\frac{1}{n}}}^{\infty}\frac{dt}{t(t^2-1)lnt}$$

(A.60)

and

$$\psi(x) = x - \ln(2\pi) - \frac{1}{2}\ln\left(1 - \frac{1}{x^2}\right) - \sum_{\zeta(\varrho)=0}\frac{x^\varrho}{\varrho},$$

(A.61)

we can clearly view in the former approximations of the oscillatory terms (A.59), the superposition of the sine functions with the frequencies γ_k which resemble in both cases a Fourier transformation of the prime number step function. Corresponding graphs are available in the vast literature on the subject.

Acknowledgements

I wish to express my sincere gratitude to the librarians at the "Handschriften-abteilung" (Department of Handwritten Documents) at Göttingen University for giving me access to Riemann's original handwritten manuscripts, in particular to the originals concerning prime numbers.

© The Author(s), under exclusive license to Springer Nature Switzerland AG 2021   105
W. Dittrich, *Reassessing Riemann's Paper*,
SpringerBriefs in History of Science and Technology,
https://doi.org/10.1007/978-3-030-61049-4

References

1. B. Riemann, Über die Anzahl der Primzahlen unter einer gegebenen Grösse. Monatsberichte der Berliner Akademie 671–680 (1859)
2. M. Edwards, *Riemann's Zeta Function* (Dover Publications, New York, 2001)
3. D. Laugwitz, *Bernhard Riemann 1826–1866* (Birkhauser Verlag, 1996)
4. R. Ayoub, Am. Math. Monthly **81**(10), 1067 (1974)
5. J. Havil, *Gamma* (Princeton University Press, 2003)
6. J. Derbyshire, *Prime Obsession* (Penguin, New York, 2014)
7. J. Stillwell, *Mathematics and Its History* (Springer, 2002)
8. G.W. Gibbons, Phys. Lett. **60A**, 385 (1977)
9. G.W. Hawking, Com. Math. Phys. **55**, 133 (1977)
10. W. Dittrich, M. Reuter, *Effective Lagrangians in QED, Lecture Notes in Physics*, vol. 220 (Springer, 1985)
11. W. Dittrich, M. Reuter, Effective QCD-Lagrangian with Z-function regularization. Phys. Lettt. **128B**(5), 321 (1983).
12. W. Dittrich, M. Reuter, Regularization schemes for the Casimir effect. Eur. J. Phys. **6**, 33 (1985)
13. E.C. Tichmarsh, D.R. Heath-Brown, *The Theory of the Riemann Zeta Function*, 2nd edn. (Oxford University Press, Oxford, England, 1986)

© The Author(s), under exclusive license to Springer Nature Switzerland AG 2021
W. Dittrich, *Reassessing Riemann's Paper*,
SpringerBriefs in History of Science and Technology,
https://doi.org/10.1007/978-3-030-61049-4

Printed in the United States
By Bookmasters